T0073055

Springer Biographies

More information about this series at http://www.springer.com/series/13617

Konrad Kleinknecht

Einstein and Heisenberg

The Controversy Over Quantum Physics

Konrad Kleinknecht
Universität München, Universität Mainz
München, Germany

Based on a translation from the German language edition:
Einstein und Heisenberg – Begründer der Modernen Physik by Konrad Kleinknecht

ISSN 2365-0613 ISSN 2365-0621 (electronic)
Springer Biographies
ISBN 978-3-030-05263-8 ISBN 978-3-030-05264-5 (eBook)
https://doi.org/10.1007/978-3-030-05264-5

Library of Congress Control Number: 2018965209

This Springer imprint is published by the registered company Springer Nature Switzerland AG
The registered company address is: Gewerbestrasse 11, 6330 Cham, Switzerland

Preface

The physics of the twentieth century rests on two foundational pillars. At the beginning of the century, our place in the universe, the origin and evolution of the cosmos, and the meaning of space and time were consolidated by Albert Einstein into a new and revolutionary picture in a relativity theory described mathematically. Thereby, he predicted a great many heretofore unknown cosmic phenomena that, in the course of time, have been discovered empirically: deflection of light in a gravitational field, black holes, stretching of time in high-velocity objects, gravitational waves, and others. Shortly thereafter, in subjecting classical physics similarly to a revolutionary transformation, Werner Heisenberg was able to explain the behavior of the smallest building blocks of matter. With his quantum mechanics, he opened up for us the world of the submicroscopic constituents of matter, atoms, atomic nuclei, and elementary particles. It also permitted description of the physical attributes of molecules, chemical bonds, crystals, solid state, and semiconductors and is thus the foundation of modern computer technology. Heisenberg's discovery of the uncertainty principle has far-reaching consequences for the philosophy of nature and epistemology.

These two great scholars both grew up in Munich and attended school there, and both were lovers of music. Along with their commonalities, though, there were also significant differences in their modes of thought. Einstein held that a physical theory must predict events precisely according to the rules of causality. By contrast, from the phenomena in the atomic domain, Heisenberg concluded that a theory can only describe possible processes and their probabilities.

Einstein left us no autobiography; he felt such books were products either of narcissism or expressions of hostility toward his fellow man. So, we must restrict ourselves in understanding his life to his correspondence and to biographies. Particularly authentic in this regard are the descriptions by his friend Philipp Frank, written in German between 1939 and 1941 in the United States. Since Einstein himself contributed a foreword to this book in 1942, it may be regarded as authorized. The Einstein estate is held at the Hebrew University in Jerusalem, and since 1987, his collected works in several volumes have been published by Princeton University Press.

Heisenberg, on the other hand, has given us a fascinating narrative of his life in his *Physics and Beyond*, which also describes his scientific breakthroughs. Additionally, two volumes of his letters to his parents and to his wife have been published. Through the agency of the Heisenberg Society, the balance of his papers has been transferred to the Archive of the Max Planck Society in Berlin; the scientific correspondence with his friend Wolfgang Pauli resides in the Pauli Archive in Geneva. His scientific papers, as well as his more generally accessible writings, are available in *The Complete Works*, published by Springer and Piper.

I wish to thank Ms. Barbara Blum-Heisenberg for providing the illustrations relevant to Werner Heisenberg and for conversations about his relation to music. Professor Hans A. Kastrup drew my attention to Albert Einstein's letter to the writer and philosopher Eric Gutkind concerning religion.

For the translation of the German book to English, I am indebted very much to Tom Artin, New York. Thanks to Maury Solomon and Hannah Kaufman for editing the book and for their ever-friendly and constructive collaboration.

München, Germany
August, 2018

Konrad Kleinknecht

Contents

About the Author

Konrad Kleinknecht is Professor of Experimental Physics. He has done research at CERN, in Geneva; at the Universities of Heidelberg, Dortmund, Mainz, and Munich; at Caltech in Pasadena; and at the Fermi National Accelerator Laboratory near Chicago. In 1988/1989, he gave the Morris Loeb lectures at Harvard University. His work on the physics of elementary particles has been recognized by numerous prizes, among them the Gottfried Wilhelm Leibniz Prize of the DFG (German Research Foundation), the High Energy Prize of the European Physical Society, and the Stern-Gerlach Medal of the German Physical Society (DPG). In addition to research papers, he has published books on the asymmetry of matter and antimatter, on particle detectors, and on the politics of German energy policy.

1

Einstein's Youth

The Cemetery at Buchau

The Jewish cemetery of the former free imperial city of Buchau in the Duchy of Württemberg lies among tall, old trees. Since 1659, the Jews of the city and the surrounding communities of the Upper Swabia have buried their dead here. More than 800 headstones, or *mazewot*, are found here. Lettering on the oldest is weathered away. Those from the eighteenth century and later are quite legible; the most recent burial took place in 2003.

Along with the neighboring Laupheim, Buchau was one of the few free imperial cities in which from the seventeenth century onwards Jews could live, and it was socially liberal. For this reason, many Jews from the surrounding area moved to Buchau. Until 1760, the community had no synagogue. In 1828 the Jews were granted Württemberg citizenship, with all its rights and obligations. In 1838, they represented one-third of the population of ca. 2,000 people in Buchau, making it the second largest Jewish community in Württemberg.

In 1838, with financial assistance from Württemberg's King Wilhelm and Prince Maximilian of Thurn and Taxis, a new synagogue was built. It became well known throughout Germany because it was unique in possessing a bell tower patterned after neighboring Baroque churches, such as the Catholic Pilgrimage Church at Steinhausen.

The first Buchau citizen with the name Einstein was Baruch Moses Ainstein, who came to the city in 1665. At the cemetery, dozens of headstone inscriptions memorialize the 99 members of the Einstein family that are buried here.

K. Kleinknecht, *Einstein and Heisenberg*, Springer Biographies,
https://doi.org/10.1007/978-3-030-05264-5_1

Fig. 1.1 Headstones at the Jewish cemetery of Buchau, © Konrad Kleinknecht

The second mayor after 1946, Siegbert Einstein, a great-nephew of Albert Einstein, is also interred here.

Einstein's father, Hermann, was born in Buchau, one of seven children of Abraham and Helene Einstein. After completing school in 1869, he and his brothers moved to Ulm. He dealt in bedding feathers and later carried on this trade together with his partners Israel and Levi in Ulm. In August 1876, in Cannstatt, he married his wife, Pauline, the daughter of the grain merchant and royal Württemberg court purveyor, Julius Koch. Following their marriage, the couple moved to Bahnhofstraße in Ulm.

The Family in Ulm and Munich

Albert Einstein was born on Bahnhofstraße, in Ulm on March 14, 1879. His mother noted anxiously that the back of his head was large and angular. He did not begin to speak until he was two and a half years old. In a nursery of today, he would count as strange. Thirty-three years later, his reticence had turned into its opposite; the physicist Max von Laue warned a colleague before his first encounter with Einstein with the words, "Watch out that Einstein doesn't talk you to death. For he likes to do that."

His father Hermann was of a contemplative nature, a good-hearted man who could not turn away any request, but was also not very capable in business. His mother, Pauline (née Koch), from a well-to-do family in Cannstatt, near Stuttgart, was witty, musical, and was an excellent pianist.

Fig. 1.2 Hermann and Pauline Einstein

In November 1881, two and a half years after Albert's birth, his sister was born, Maria (called Maja), with whom he had a lifelong, warm relationship. Later, in a biography, she described experiences from her childhood. She noted particularly the inexhaustible patience with which her brother worked alone on his "projects." He built palaces and castles from his Anchor Stone blocks, worked out figures from plywood with his jigsaw, and built tall, precarious buildings from playing cards. Drilling through thick boards was later one of his trademarks in physics.

Fig. 1.3 Albert Einstein with his sister Maria (Maja), 1885

Searching for special mathematical and physical gifts among Albert Einstein's relatives, one finds his Uncle Jakob (1850-1912). Hermann Einstein's younger brother had studied electrical engineering in Stuttgart at the Polytechnic Institute, and there learned the "Maxwell Equations," newly discovered by James Maxwell. They were formulated in final form in 1864. In the 1870 war, Jakob served as engineering officer. Following the war, he decided to employ his expertise to found a company in Munich for the manufacture of direct current generators and electro-motors. He designed his machines himself and had them built in the workshop.

Jakob suggested his brother Hermann join the firm and take up the position of commercial executive. Hermann agreed to the proposal and moved to Munich in 1880, at first to Müllerstraße 3, where Jakob had his apartment and his shop. The electro-technical factory J. Einstein & Co. offered the "construction of electrical power transmission plants," as well as the "construction of electric lighting systems, fabrication of generators for lighting, power transmission, and electrolysis" and enjoyed success. The International Electricity Exhibit, organized in 1882 in the Glass Palace in Munich by Oskar von Miller, future founder of the German Technical Museum (dubbed "Deutsches Museum"), brought the new technology into the limelight. The firm Einstein & Co. exhibited their generators as well as a telephone exchange.

In 1885, the Einsteins purchased a new business site on Lindwurmstraße. They lived at Adlzreiterstraße 14, which today bears a plaque commemorating their residence. The whole extended family was gathered in this house – Hermann and Pauline, together with Albert and Maria on the second floor; Uncle Jakob, his wife Ida and Pauline's father, Julius Koch, on the ground floor. The two men and Ida ate with Albert's family, and naturally Uncle Jakob spoke about his area of expertise, electro-dynamics and its applications. Albert was probably the only 15-year-old student in Germany for whom the Maxwell equations were regular topics of meal-time conversation. In these equations the letter *c*, referring to the speed of light, occurs; this must have made an impression on him even then. Albert's major interest was in mathematics, in which it is possible to prove the correctness of statements oneself. A small pamphlet containing the axioms of Euclidean geometry was sacred to him. Another critical experience was his father's demonstration of a compass. The force that turned the compass needle in the northerly direction fascinated the boy. He wished to understand this mysterious phenomenon.

Initially, however, in 1885, Albert attended the Catholic St. Peter's Elementary School, where a strict regimen prevailed. He was not pleased with the drills; he disliked answering questions by rote and wished to think things through for himself. He was first in his class, and his intelligence earned him respect. As the only Jewish member of the class, he took part in the Catholic religious training, learning the biblical stories of the Old and New Testaments.

Fig. 1.4 Dwelling of the Einstein family on Adlzreiterstraße in Munich, © Konrad Kleinknecht

Student at the Luitpold-Gymnasium, Munich

In October 1888, Albert entered the Luitpold Gymnasium. Among his classmates were Robert Kaulbach, a member of the well-known family of painters, and Paul Marc, the older brother of Franz Marc. With the founding of the Blue Rider movement, Franz Marc, along with Wassily Kandinsky, became a revolutionary in painting, as was Einstein in physics. Einstein was an outstanding student at the humanistic Luitpold Gymnasium, excelling especially in mathematics.

Fig. 1.5 Gymnasium student Albert Einstein, age 14 years, in Munich

All sorts of authority were an anathema to him. He hated learning Greek and Latin vocabulary mechanically by rote. When the essence of the ancient culture was conveyed by way of the language, however, he was enthusiastic. He was most impressed by his teacher Rueß, who had the knack of presenting the ancient ideas and their influence on German culture in a lively manner. No doubt the conceptions of the Greek philosophers about nature, and their speculations about symmetries and mathematical laws, were topics of discussion that fit with Einstein's artistic nature. Accordingly, Albert always received good, even excellent, grades in the ancient languages. His teacher Rueß also taught German literature, of which Einstein's most vivid memory remained his reading of Goethe's *Hermann and Dorothea*. Schiller's dramas, too, with their idealistic heroes, had their charm for him.

Einstein's disdain for all forms of authority led to a tense relationship with several of his teachers. He felt compelled to let his teachers know his intellectual superiority. Later, during his studies at the Federal Polytechnic in Zürich, he behaved similarly. One of his professors there said to him, "You are a clever young man, Einstein, a very clever young man. But you have one great shortcoming. You don't listen to anyone else."

His skepticism regarding authority led also to the insight that, on closer scrutiny, the religious truths in the Bible "could not be consistent" with the broad context of natural science. At the Bavarian Gymnasium, religious instruction was mandatory; there was the academic subject "Israelite Religious Education," of which he partook. In this instance, he was not merely a disinterested auditor, as he had been in Catholic religious instruction in elementary school, but rather a regular participant. Though Einstein's parents did not observe the traditions of Judaism, here he was introduced to the Talmud and the Old Testament, as earlier in elementary school to the New Testament. Students were naturally required to participate in religious services in the synagogue. Einstein experienced this as coercion and formalistic routine.

At the age of twelve Albert read books such as the *Popular Natural Scientific Books* by Aaron Bernstein. Reading these, he became aware of the contradiction between the biblical stories and science. He became a free-thinker. The conclusion he drew from this was that if youth is intentionally lied to in its religious instruction, then perhaps the "truths" in school books might be false as well. His suspicion of all forms of authority was thus confirmed. He even contemplated withdrawing from the Jewish religious community following gymnasium studies, although it was only later that he actually took this step.

His Uncle Jakob, who lived with the family in the same house, exercised considerable influence on Einstein. He set Albert mathematical problems, convinced that they were too difficult for the boy. Naturally, Albert solved them nevertheless. When his uncle referred to the Pythagorean theorem, the 12-year old Albert determined on finding a proof. It took him three weeks, but he stuck to it until he had found a solution.

In music, too, he applied this patient energy as soon as the substance of pieces took hold of him. Although in the early years of violin playing the technical requirements for mastery of the instrument first have to be acquired, the practices are often boring and musically unsatisfying. For this, Albert cared little. But as soon as he became aware of the great works, his interest grew, and he made an effort to acquire the technical skills for him to play his beloved Mozart violin sonatas. He retained his love of music all his life.

Meanwhile, the electrical firm J. Einstein & Co. prospered. From time to time, Albert visited the factory and thus acquainted himself with the application of the theory of electromagnetism. When he learned of a problem with production over which his Uncle Jakob had ruminated unsuccessfully for days, he quickly found the solution, much to his uncle's delight.

With an eye to publicity, the Einsteins laid an electrical cable from their factory on the Lindwurmstraße to the Theresienwiese for the Octoberfest of 1885. The Octoberfest tents were illuminated with current from the Einstein generators together with oil lamps. Because of a fire caused by one of these lamps at the Octoberfest of 1887, in 1888 illumination of the tents was switched entirely to electricity. The Einstein firm received the contract. The same year, the conversion from gas to electricity for the street lighting of the Munich district of Schwabing was announced, and once more the Einstein firm was awarded the contract. The new illumination was inaugurated in February 1889 to great ceremony. The celebration concluded with fireworks, rockets, and cannon fire. Jakob Einstein donated the lighting installation to the city of Munich.

At this time, the Einstein firm employed 200 workers; the family became well-to-do. But already the following year, powerful competitors came on the scene, among them Schuckert & Co., in Nürnberg, AEG, and Siemens & Halske, who all used alternating current technology. In 1892, the entire Munich street lighting system was up for grabs; all competitors submitted bids. The best offer, from Schuckert, won the contract. The bid from J. Einstein & Co. was higher by far.

In the wake of this failure, the Einstein firm had to lay off many employees; their competitors took over the lucrative commissions. In the summer of 1894, Hermann and Jakob Einstein decided to liquidate their firm and start over in Italy, where relatives of the family lived. They opened a similar firm in Pavia.

After his parents had moved to Italy, Albert remained alone in Munich, and was supposed to complete the *abitur* (secondary school exams), which were prerequisite to university study. In the fall, he entered the 7th grade (today, the 11th grade) of the gymnasium. Because he was at odds with the teacher, and the teaching methods used were unbearable to him, he was gradually confirmed in his decision to leave school. One motive could also have been that it was more difficult after the age of 16 to give up one's German or Württemberg citizenship, and thus avoid military service. From a friendly doctor, Albert got certification that he had suffered a nervous breakdown. Accordingly, a six-month's period of recuperation with his parents in Italy was medically advised. Since he knew that he would need a degree, he received from his mathematics teacher a document in writing that attested to his extraordinary abilities in mathematics that qualified him for admission to another gymnasium. His exit from the Luitpold Gymnasium was then surprisingly easy, because in December 1894 his behavior provoked quite a stir; his teacher asked Albert to leave the school because his mere presence in the class undermined respect for him. On December 29, 1894, he left the school and traveled to join his parents in Italy.

Einstein in Aarau and Zürich

After his flight from Munich in December 1894, Einstein traveled to his family, and for half a year enjoyed life in Italy. He explained to his father that he wished to relinquish his Württemberg citizenship, and that he had withdrawn from the Jewish religious community. In the summer of 1895, he had no firm idea how things would proceed. In the meantime, he considered joining his father's firm. He gave up this idea because neither in Pavia nor in Milan were his father's enterprises prospering. It was now clear to him that he had to plan his professional future.

He hoped to be able to begin study at the Swiss Polytechnic (called "Polytechnikum") Institute in Zürich, which was among the best technical schools in Europe. His father spoke to a friend who lived in Zürich, who turned personally to the director of the Polytechnic Institute. The director permitted Albert to take the admission exam for study towards a teaching post in mathematics and physics. In the examination, Albert excelled in mathematical and scientific subjects, but his knowledge of modern languages, literary history, zoology, and botany was insufficient. He was not accepted.

The director of the Polytechnic Institute, Albin Herzog, advised Albert that even prodigies had to obtain the abitur, and he recommended the Canton school in Aarau. This gymnasium was well equipped with physics and chemical laboratories, a zoological collection, and geographical visual aids. A microscope was also available. Einstein was fortunate enough to be taken in as a lodger in the house of a teacher, Professor Jost Winteler, who taught Greek and history. Winteler looked after him, and took him and his two children on outings into the mountains. Einstein had many discussions with him about politics in democratic Switzerland and how it compared to that in Germany with its emperor Wilhelm II.

In just under a year, Einstein completed the abitur at the Canton school. In September 1896, following the written part of the graduation exam, came the oral portion. In an essay in French, "*Mes projets d'avenir*" (My Future Projects), the 17-year-old Einstein announced his plans to study physics and mathematics and become a professor of theoretical physics. So during his time in Aarau, he veered from a career path in engineering – following in his Uncle Jakob's footsteps, as his father wished – revealing his true interest in the theoretical study of nature.

Some of his graduation grades were the best in his class. In both mathematical subjects, geometry and algebra, he received the highest grade, a 6, in physics a 5-6. The grading system, which differed from that in Germany, later led to confusion in German reports that Einstein had been a poor student. It may be consoling to poor students that the renowned genius had been a failure in school, but actually quite the opposite is true.

During his time in Aarau, Albert developed an especially warm relationship to Marie, the daughter of his teacher Winteler. She was two years his senior and loved Albert passionately. But as soon as Einstein had the Swiss graduation diploma in his pocket, his goal was the Polytechnic Institute in Zürich. He commenced his studies there in October 1896; he had to focus on that and had no time for distractions. His studies commanded his whole mental effort. He fled from potential romantic complications and avoided further visits to Aarau. Connection to the Winteler family continued, however; Albert's sister Maja later married the youngest Winteler son Paul.

The Polytechnic in Zürich was the only school of higher education financed by the Swiss federal government, in contrast to the universities supported by the Cantons. To the present day, the ETH (*Eidgenössische Technische Hochschule*) occupies the magnificent building designed by Gottfried Semper of Dresden. Einstein signed up for the course of study for "teacher of mathematical and scientific subjects." Among the dozen students who shared this interest was a young Serbian woman, Mileva Maric'.

Fig. 1.6 Mileva Maric' in Zürich, 1900

Einstein was very interested in Professor Heinrich Friedrich Weber's lectures on thermodynamics. On the other hand, he was uninterested in the course called Experimental Physics Laboratory for Beginners, which consisted of practical experimental training. In this subject he was reprimanded "for lack of diligence," and received the lowest grade, a 1. He neglected mathematical instruction, too, because he considered his knowledge sufficient. As a result, in formulating his theoretical works later, he often had to seek the help of mathematicians.

In the course of his studies, Mileva soon became more than an empathetic fellow student. She shared his scientific interests, and so he broke contact with Marie Winteler in Aarau. Mileva studied for one semester in Heidelberg. In his letters, he encouraged her to return. He was pleased about her intention to pursue her studies here, and he encouraged her to come to Zürich as soon as possible. After her return, they continued with their private studies. Only in the final student year did they use the familiar "du." He called her Doxerl; himself, Johonserl. Both diploma theses written under Professor Weber carried the title "Thermal Conduction," and were "of no interest" to Albert. Einstein received an average examination grade of 4.91 out of 6, or "sufficient." Mileva received a 2.5, or "insufficient," for one of the subjects in the written exam, Complex Analysis, and did not receive a diploma. She had to repeat the examination the following year, but once again failed to pass.

Expert at the Bern Patent Office

Following his teacher's exam at the Polytechnic in Zürich in July 1900, Einstein hoped for an assistantship with his professor, Heinrich Friedrich Weber. He wanted to marry Mileva, so the question of his livelihood was pertinent. He was aiming for a career as secondary school teacher in Switzerland, and he assumed Swiss citizenship would be an advantage to him in this. Having given up both his Württemberg and his German citizenship he was technically stateless. So he saved up the 600 francs required to apply for Swiss citizenship. The immigration commission of the city of Zürich was primarily interested in a candidate's financial circumstances. In February 1901 he was able to give evidence of this sum, and had the approval of the Canton officials; at that time he became a citizen of the city of Zürich, and of Switzerland. Once a citizen, he was called up for military service, but due to his varicose veins, and flat and sweating feet, he was declared unfit to serve. So he managed to avoid the military discipline he had flown Germany to evade. In this way, he could also hold onto his vision of a democratic Switzerland that Winteler, his teacher in Aarau, had instilled in him. Nevertheless, his spirit and his speech remained thoroughly German.

Albert did not receive the assistantship to Professor Weber he had hoped for, nor were the applications he sent to many European institutes successful. His applications to Friedrich Wilhelm Ostwald at Leipzig, and Heike Kamerlingh Onnes at Leiden, went unanswered. To that point, he was, to be sure, totally unknown in the academic world, and had not yet even earned his doctorate. (Remarkably, Ostwald was, in 1910, the first to nominate Einstein for a Nobel Prize.) So Einstein had to seek teaching positions.

He got his first, short-term position in 1901 as an assistant teacher in a gymnasium in Winterthur. That fall, an opportunity arose to take a position as private tutor assisting a mathematics teacher at a gymnasium in Schaffhausen on the Rhine. He was to tutor a student in preparation for high school graduation. This position left him time enough to write a dissertation on kinetic gas theory, which he submitted to the University of Zürich. It was not accepted. He could not submit the paper to the Polytechnic because this institution was not yet authorized to award a doctorate.

During this time, his friend Mileva remained in Zürich, but in July 1901 she returned to her parents' home in Novi Sad, Serbia, which was part of the Austro-Hungarian Empire. She was pregnant by Albert, and had failed in her second attempt at the diploma exam. She also knew that Albert's parents rejected her as a daughter-in-law. They even wrote Mileva's parents, telling them that they did not approve of the marriage. In October, Mileva visited Albert in Schaffhausen, though she lodged discretely at a hotel in Stein on the Rhine. Two weeks later, she traveled back to Novi Sad, and in January 1902 gave birth to their daughter.

Albert wrote to her in February and asked after the child. Meanwhile, from his friend from student days, Marcel Grossman, he had learned that he could apply for a civil service post with the Swiss Department of Intellectual Property at the Patent Office in Bern. Grossmann arranged a meeting with the director of the office, Friedrich Haller. Following this, Einstein sent in his application and was quite confident that he would receive the position. He quit his job in Schaffhausen "with a bang" and moved to Bern. But he did not want Mileva to bring the child with her when she returned to Switzerland; he presumably feared that an illegitimate child would compromise his application for the position at the patent office. Additionally, under civil law then prevailing in Zürich, it was possible to recognize the illegitimate child only through a formal adoption process, which would have drawn attention. So the child, who was called Lieserl, remained in Novi Sad.

From the summer of 1901 on, there are no letters preserved that mention her. Her name was systematically suppressed. For the rest of his life, Einstein never spoke of her. Her further fate is in dispute. She may have been sent to Germany and adopted there by a couple by the name of Gießler, living until 1980 as Marta Zolg (née Gießler) in Bietingen, near Konstanz.

Finally, in June 1902, Einstein's appointment at the patent office was confirmed by the Federal Council. The civil service job at the Patent Office suited Einstein much better than working as a school teacher because his work as a patent examiner left him a lot of free time for his own research interests. The work at the patent office also came easy to him because during his student years in Munich he had learned the technical details of electromagnetic generators and motors at his father's and uncle's company, Einstein & Co. Einstein judged the work of assessing patent applications as uncommonly diverse; it gave him much to ponder. In one testimonial that survives, he rejects an application for a patent from AEG, Berlin, for an alternating current commutating machine because in his opinion the title was inaccurate and imprecise.

Fig. 1.7 Einstein at the patent office, Bern, 1905

Marrying Mileva proved difficult. Einstein's mother, Pauline, wrote to a friend, "Were it in my power, I would do everything possible to exclude her from our circle. I feel a palpable antipathy to her." His father, Hermann, was also opposed to the marriage. He finally assented to the marriage only in the fall of 1902 in Milan on his deathbed. The marriage took place in January 1903 without either family in attendance. Through the marriage, Mileva became a Swiss citizen, whereas for the time being their daughter "Lieserl" (or Marta) remained a citizen of Austria-Hungary. In 1904, their elder son, Hans Albert was born in Bern. Then, in 1910, in Zürich, a second son, Eduard, was born.

As their son Hans Albert later indicated to a journalist, the exclusion of their sister from their life (no doubt at Albert's instigation) weighed upon the marriage from the start. Though Hans Albert knew nothing of his sister, his mother had told him that something had transpired between the two for which Albert was to blame.

Even before taking the position at the patent office, Einstein had found two friends in Bern with whom he could read and discuss philosophical, scientific, and literary texts. One was Maurice Solovine, a Rumanian student of philosophy; the other, Conrad Habicht, was a Swiss student of mathematics. The three founded the Akademie Olympia, a reading group that met regularly for discussions over tea and sandwiches. They read philosophical works that treated scientific topics such as those by David Hume, Immanuel Kant, Ernst Mach, and Henri Poincaré, but also "for edification" purely philosophical writings by Arthur Schopenhauer and Friedrich Nietzsche. Einstein's view of women could be said to have been marked by Schopenhauer, who characterized women as "life-long grown-up children," whereas a man was "the essential human being."

From David Hume, who represented the English Enlightenment, Einstein espoused the concept that scientific knowledge rests on empirical experience, and that it can be described mathematically. Hume criticized the method of induction, by which general laws are based on individual cases. He taught that observation only shows which processes regularly occur if certain conditions are met, but that beyond that we cannot derive a causation.

The philosopher Ernst Mach also impressed the reading group Akademie Olympia with his positivism. He placed several fundamental concepts of Newtonian mechanics on the testing bench and found that there is no empirical basis for concepts such "absolute space" and "absolute time." Here we have a starting point for Einstein's later insight that all frames of reference in uniform motion relative to each other are equally valid.

Einstein accepted the role assigned by Immanuel Kant to human reason. However, he rejected the Kantian idea that reason can know laws that are "a priori" and valid for all time. Clearly missing here was reference to empiricism, which the positivists regarded as fundamental to knowledge of nature.

Fig. 1.8 Albert and Mileva Einstein in Bern, 1905

The time at the patent office in Bern was a happy time for Einstein. Mileva cared for the little family with their son, Hans Albert; Einstein derived satisfaction in his work as a patent tester, which enabled a carefree life. A practical occupation, he wrote later in retrospect, is a blessing for people of his kind. Aside from their professional work, people with deeper scientific interests could immerse themselves in their favorite problems. Einstein used the free time he had outside his work at the patent office to overturn single-handedly the world of physics.

2

Heisenberg's Youth

Heisenberg's Background

Heisenberg's parents came from quite disparate backgrounds. His grandfather, Wilhelm August Heisenberg, came from a Westphalian family of artisans. His ancestors were distillers of spirits and coopers in Detmold and Osnabrück. He himself ran a locksmith's shop in Osnabrück and devoted himself to assisting the poor.

August Heisenberg, Werner's father, was born in Osnabrück in 1869. He attended the public grammar school and the high school, and then broke with the tradition of the artisan family in that from 1888, he studied at the University of Marburg. At first he vacillated between philosophy and theology, until the theologian Adolf von Harnack advised him to study philosophy. Because of his love of music, he transferred to Munich University, where through his acquaintance with Karl Krumbacher, an interest in ancient Greek culture was awakened.

After his teacher's exam, he began his pedagogical training at the Maximilians-Gymnasium, where he met his future father-in-law, Nikolaus Wecklein. After his assistantship at Landau in the Bavarian Palatinate, in military service in Osnabrück, and as a teacher at the Maximilians-Gymnasium in Munich, he took up his first student teaching position in Lindau on Lake Constance. He had previously become engaged to Wecklein's eldest daughter. He must have been a very lively, enterprising young man, who enjoyed teaching and who, with his pedagogical gifts, enjoyed great success with his students. But his scientific interests moved him to apply for the Bavarian State Archaeological scholarship. He won it and spent 1898 and 1899 in Italy and

K. Kleinknecht, *Einstein and Heisenberg*, Springer Biographies,
https://doi.org/10.1007/978-3-030-05264-5_2

Greece. He returned to the Luitpold Gymnasium in Munich with the resolve "to devote his efforts from then on to researching ancient and modern Greek culture."

In 1899, he married Anna Wecklein. Their son Erwin was born in March 1900; Edwin's younger brother, Werner, was born in December 1901. In 1901, August Heisenberg was transferred to a gymnasium in Würzburg. There he taught Latin, German literature, and geography. At the same time, he did further research in his second field, Byzantine culture, gave lectures at the university, and published 50 scholarly papers.

In his work at the school, he was assisted by his wife, Anna. She learned Russian in order to help her husband with translations of Russian sources on Byzantine culture. Annie (as she was called in the family) focused her attentions on their younger son, Werner. The father favored the elder boy, Erwin. Werner inherited his tireless creativity and optimism from his father.

August Heisenberg did research on Byzantine culture and art in Turkey. He was ultimately appointed to the professorship in Byzantine culture at Munich. He died in 1930 of malaria, contracted on his travels.

While August Heisenberg is described as temperamental, his wife Annie was rather a calm, poised person. The maternal Wecklein family included merchants, farmers, clerics, artists, and academics. Among them was the violin virtuoso August Zeising, whose son took part in the 1848 revolution and was later a member of the Bavarian Academy of Sciences. His daughter Magdalena married the philologist of ancient language Nikolaus Wecklein (born, 1843). Wecklein received his doctorate on the Greek Sophists in 1865. After a few years of teaching at the Ludwigs and the Maximilians-Gymnasium in Munich, in 1869 he completed his dissertation on Greek inscriptions. Following several intermediate positions in Bamberg and Passau, he ultimately became Rector of the Maximilians-Gymnasium in Munich, and a privy councilor. His grandsons, Erwin and Werner Heisenberg, were students in his school growing up.

Heisenberg's mother, Annie, lived as a widow in Munich, and, during WWII, mostly in Urfeld. She was a clever, educated woman, even though she had not studied formally. She is said to have corrected the Latin and Greek papers of her husband's students, and even closely followed her husband's work. She wrote poetry and made up stories for her grandchildren. Her relationship to her son Werner remained close all her life. She died in 1946 in Bad Tölz.

Werner maintained that he had inherited his calmer temperament from his mother. But he no doubt had his father to thank for his "Westphalian block-headedness." Years later, his wife, Elisabeth, often joked about him with the saying, "don't bother to honk – I'm Westphalian."

Schooldays in Würzburg and Munich

On weekends, the Heisenberg family liked to hike through the vineyards and woods around Würzburg, including making stops at local inns. They also visited grandparents in Osnabrück and Munich. In the fall of 1907, Werner entered the grammar school in Würzburg. Only one event is documented. After being caned so severely by his first grade teacher that his hands were swollen, he refused to work with that teacher again and withdrew. This reaction to what he considered unjustified punishment remained a character trait of Werner's in his later life as well.

Three years later, the family moved to Munich because the father, August, accepted a professorship in Byzantine studies at the University of Munich, which the philo-hellenistic King of Bavaria had established there. This meant a transition from a small city in a rural environment to the Bavarian capital, which was starting to emerge as a metropolis. The woods and the hilly hiking country were gone. Werner felt his confinement in the city apartment as a loss.

Fig. 2.1 The family of August and Annie Heisenberg, with their sons Erwin and Werner, on a hike near Würzburg, 1907, © Werner Heisenberg Estate

Fig. 2.2 Werner (left) and Erwin visiting their grandfather in Osnabrück, 1906, © Werner Heisenberg Estate

Since both brothers were ambitious and competitive, Werner's relationship to Erwin was tense. Their father steered their rivalry into intellectual pursuits by challenging the boys to compete in solving mathematical problems. Werner noticed that he was quicker than his older brother in this discipline and developed from then on a particular interest in mathematics. Their father also encouraged the musical training of both sons. Erwin learned the violin, Werner at first studied the cello, then the piano. Already at 13, Werner could sight-read, accompanying his father on songs, and could take the piano part in chamber music played in the house. Piano trios and sonatas were often on the program. Werner even considered becoming a professional musician.

Fig. 2.3 Nikolaus Wecklein with his grandsons, Erwin and Werner, 1915, © Werner Heisenberg Estate

In the fall of 1911, one year after his brother, Werner entered the Maximilians-Gymnasium in Munich, whose director was his grandfather Wecklein. From the outset, he was an excellent student. One report reads, "He has achieved his considerable accomplishments with playful ease; they have cost him no great effort…He is also appropriately self-assured, and seeks always to excel."

Particularly noted already in the third grade of Gymnasium (corresponding to seventh grade today) was his remarkable knowledge of physics. His passion seemed to involve working with and building small machines.

An especially enlightening experience for Werner was the idea of the mathematician Wolff that it is possible to arrive at generally valid statements about geometrical forms such as triangles and rectangles, and "that one [can] recognize and assess certain conclusions not only from figures, but also prove them mathematically."

I found this idea, that mathematics somehow fits with the forms of our experience, extremely remarkable. I tried the application of mathematics for myself, and I found this play between mathematics and direct experience to be at least as interesting as most other games…. Later, from Göschen volumes and similar somewhat primitive textbooks, I began to learn the mathematics one needs to describe laws of physics, primarily differential and integral calculus.

At Werner's request, his father borrowed from the university library Leopold Kronecker's dissertation on number theory – in Latin. Werner wrote a short paper on this work.

We see here the parallels between Heisenberg's private extracurricular study and Einstein's "little holy geometry book," his conversations with his Uncle Jakob, the engineer, and his readings in books on popular science.

The outbreak of WWI interrupted this idyllic period. Heisenberg writes, "When my father entered our room with news of the declaration of war, I inferred from my parents' expressions that a misfortune of the worst sort had occurred, that would affect not just us, but all mankind." For four months during the last year of the war, Werner worked with a group of his contemporaries at a farm near Miesbach, haying and chopping wood.

The new school year began with the armistice of the Imperial troops, and the abdication of the emperor on November 9, 1918. In Bavaria, after the murder of the independent socialist Kurt Eisner, the left established a soviet republic, after the Russian model. The "red terror" was soon followed by the "white terror," when the Army of the Reich and Knight Franz von Epp's Freikorps seized the city. Werner served for several weeks in a cavalry unit, which had quartered in a seminary across from the university.

In order to prepare himself once more for school, Werner withdrew to the roof of the seminary with the Greek edition of the Platonic dialogues and read the Timaios, which discusses the smallest particles of matter. It was Plato's idea that the smallest particles were regular geometrical bodies. The four elements were composed of such bodies – the earth of cubes, fire of tetrahedrons, air of octahedrons, and water of icosahedra. However, it seemed quite absurd to young Heisenberg that just these regular geometrical bodies should be the fundamental elements. Besides, there were five of these regular symmetrical bodies, and one had to assume, with Aristotle, that corresponding to the regular dodecahedron (with twelve surfaces) there must be a fifth element, the transparent ether. Nonetheless, this idea impressed Heisenberg so much that even decades later he had small paper models of the regular Platonic bodies sitting on his desk.

At the gymnasium, now, he was working toward his graduation diploma. New subjects were required. To the ancient languages, French was added, and to mathematics, physics with the classical divisions: mechanics, electricity and magnetism, thermodynamics, and optics. The molecules illustrated in his school books, in which atoms were connected with small hooks, amused Werner. That could no more represent reality than Plato's ideas.

Heisenberg was undisputedly the top student in his class, yet he was modest and shy. This remained the case up to his graduation in the summer of 1920. The report reads, "In the fields of mathematics and physics, his knowledge significantly exceeds the school framework." But even in almost all other subjects, among them the three languages, Latin, Greek, and French, and in history and physical education he received a "very good." The only subject in which he was judged merely "good" was German language and composition.

Because of his outstanding record on graduation from high school, Heisenberg was accepted to the Bavarian Maximilianeum foundation for exceptionally gifted students.

The Youth Movement

After the end of WWI, the youth movement growing out of the Wandervogel movement and the Boy Scouts in the first third of the twentieth century attracted large segments of young people. Following the fall of the German Empire, a new generation wanted to blow up the old traditions, and create new value systems – turning toward nature in place of urban life and industrialization, music, theater, and handicrafts. Young people also wanted to try out new forms of education. Already as a student, Heisenberg had participated in Boy Scout groups; later he was active in adult education. As a sixth former and as a university student, he gave courses in astronomy and music at the adult education center. He joined the youth movement in Bavaria, the Bavarian Boy Scouts and Young Bavaria, Bavarian State Association for Cultivation of Physical Training of Youth. Heisenberg became leader of a small group that met to plan hikes and other endeavors. They slept in tents, or in the early youth hostels just then emerging.

In August 1919, Heisenberg's group traveled by rail and by foot to the first German Boy Scout Day that took place at Schloss Prunn in the Altmühl Valley. High school and university students, as well as young men returning from the war, had gathered there. The question of whether young people had the right to organize their lives themselves, according to their own values, was the subject of lively discussion. Heisenberg was especially impressed by how, following these discussions in the middle of the night, silence fell, and a violinist played the Bach *Chaconne* from the balcony of the palace courtyard.

The youth movement was a predominantly male community, in line with the gender-separate education prevalent in schools of the time. Friendships formed in such groups lasted a lifetime. But they were also the reason the members of the group married late or not at all.

The romantic ethos of the Boy Scouts was expressed in Stefan George's poem:

Whoever has marched 'round the flame,
Stay satellite to the flame!
Just as it wanders and circles:
If light of the flame still reaches,
It strays not too far from its goal.
Only when lost from sight,
Its own light deceived it:
Absent the law of the center,
Is scattered and sent into space.

The poem describes not only the mood around the campfire but also the search for the center that holds the world together. This corresponded to Heisenberg's conviction.

Studies with Sommerfeld

In the fall of 1920, after his term of service as harvest hand at the farm in Miesbach and the revolutionary unrest of the post-war period, Heisenberg began his studies at the Ludwig Maximilians University in Munich. His reading of Einstein's presentation of relativity theory, and Hermann Weyl's book *Space, Time, Matter*, occupied and unsettled him so much, that he determined to study mathematics in order to understand the mathematical methods and the underlying constructs developed there. He believed the mathematical knowledge garnered from his self-study was sufficient to allow him to attend the seminar of the famous mathematician Lindemann, just as Einstein had believed himself qualified for admission to study at the Swiss Polytechnic without a high school diploma. At a personal interview, Lindemann reacted with annoyance to the presumption of the future student; participation in the seminar was actually open only to advanced students. When Heisenberg mentioned his reading of Weyl's book, Lindemann's patience was exhausted – in that case he was already ruined for mathematics!

Heisenberg decided to change fields, and chose theoretical physics. The theoretical physicist Arnold Sommerfeld, who he interviewed with, was friendly and sympathetic and and became his teacher. Sommerfeld was recognized internationally as one of the leading scientists in the area of theory of the smallest component of the elements, the atom. Sommerfeld told the young Heisenberg that as a beginner, Weyl's book was much too difficult for him. He would do better to start with more modest, more traditional work in physics.

Fig. 2.4 Arnold Sommerfeld, 1920

During his first semester of studies, Werner encountered fellow student Wolfgang Pauli from Vienna, who was one year older and already in his fifth semester. They became close friends for their lifetimes. Beside the study of classical physics, the two were ardently interested in the current problems of relativity theory and atomic physics.

Even in their first years, both students followed the latest developments. When Heisenberg asked Pauli whether relativity theory or atomic theory was more important, Pauli expressed the opinion that the special theory of relativity was complete, and no longer of interest to anyone wishing to find something new. The general theory of relativity, on the other hand, was not yet complete; in 100 pages of theory, there was only one experiment, so it was still unclear whether it was correct. He found atomic physics fundamentally much more interesting because it included a wealth of unsolved problems.

Fig. 2.5 Wolfgang Pauli, 1918, © CERN, Geneva

Both students mastered the material of traditional classical physics with playful ease, insofar as theoretical physics was concerned. They were less engaged in attending lectures on experimental physics and carrying out the experimental work. Naturally, it was evident to Sommerfeld the level of talent he had in these students listening to him at his lectures. To be sure, only Heisenberg attended regularly; Pauli was fond of sitting in cafés and bars until late in the night, and so could not attend lectures at early hours. He asked the professor kindly not to erase the last blackboard he had used during his lecture, but to leave the formulas in place. Then he could come by around noon, and from the contents of the last blackboard, he would understand the previous material. And so it was.

Sommerfeld knew exactly how gifted Pauli was. When, as editor of the prestigious journal *The Encyclopedia of Mathematical Sciences (Encyklopädie der Mathematischen Wissenschaften)* he asked Einstein for an article summarizing his relativity theory, Einstein first agreed. Then in 1918 said he was too busy with his own research projects to write it. So Sommerfeld, with Einstein´s agreement, passed the assignment on to the 20-year-old Pauli. Pauli quickly penned an article that was created as a comprehensive presentation of the special and general theories of relativity, and in this capacity it immediately became a standard reference, frequently cited to this day. Pauli's essay from

1921 stands even now as one of the best presentations of this theory, on whose basis we understand space and time. Einstein wrote of this, full of praise:

One doesn't know at what one should marvel most: the psychological comprehension of the development of ideas, the sureness of mathematical deduction, the penetrating physical insight, the capacity for lucid, systematic presentation, the knowledge of the literature, the objective integrity, the confidence of the critique.

Pauli wrote his doctorate on a topic given to him by Sommerfeld: the ion of the hydrogen molecule.

Heisenberg worked on various problems of atomic theory, but for his dissertation, Sommerfeld had other ideas. It was his opinion that before taking up the current theme of atomic physics, a student had to solve a different problem of classical physics. Perhaps he was also taking into account that the co-examiner for experimental physics, Professor Willy Wien, took a critical position *vis à vis* the theoretical model of the atom.

In any case, Sommerfeld posed to Heisenberg the assignment of clarifying the hydrodynamic conditions of liquid flow. To do this it is necessary to calculate at what velocity a slow, uniform laminar flow in a pipe converts to a turbulent flow. Since Sommerfeld was teaching in the United States in the winter of 1922-23, he sent Heisenberg to Göttingen for this time, where he heard the physics and mathematics lectures of Max Born and Richard Courant. Heisenberg gave a lecture in Courant's seminar and worked on his dissertation. In the spring of 1923, Heisenberg returned to Munich and concentrated on the elaboration of his dissertation, on which he had previously worked only intermittently. In April, he completed the work with the title *On the Stability and Turbulence of Liquid Flows*, and submitted it to the faculty. In the paper he deals with both types of flow and shows how the transition between them can be characterized with the so-called Reynolds number.

He used his free time thereafter to better his insufficient skills as an experimenter by carrying out an experiment in the physics internship for advanced students in measuring the splitting of spectral lines of mercury vapor in a magnetic field. This phenomenon, the "Zeeman effect," was named for its discoverer.

The oral examination for the doctorate, the *Examen Rigorosum*, took place on July 23, 1923. Although his exams in mathematics and astronomy were graded "very good," and "good," the meager interest the student had evinced during his studies took its toll in the exam for experimental physics. He was simply unable to answer questions about the resolving power of the microscope, and about the function of the lead accumulator. Willy Wien was so annoyed he wanted to fail the candidate. Only Sommerfeld's forceful defense saved Heisenberg. Sommerfeld had judged the dissertation as excellent, thought Heisenberg brilliantly gifted, called him his most talented student, Pauli and Debye not excepted, and looked for "tremendous things" from him. The two physicists agreed on a grade of III, *cum laude*, and this was also his overall grade, a disheartening result for Heisenberg.

Sommerfeld had arranged for a small celebration at his house, but Heisenberg was so despondent that he took the train to Göttingen that very night. The next morning, he showed up at Max Born's and asked if after this abortive exam he still wished to have him as his assistant. Born asked for the details of Professor Wien's exam questions, but saw no reason to send Heisenberg packing. Not for a moment did he doubt his extraordinary abilities.

Before the start of the winter semester at Göttingen, in September 1923, Heisenberg attended the conference of the Society of Natural Scientists and Physicians in Leipzig, at which he hoped to hear the world-famous Einstein on relativity theory. At his entry into the conference hall, a young man pressed into his hand a card that warned against Einstein and his relativity theory. The text was by Philipp Lenard, the famous Nobel laureate from Heidelberg, who had discovered cathode rays.

Heisenberg was familiar with his friend Pauli's article on relativity theory. He knew, too, that empirical findings supported the theory. For him, Einstein's theory was a self-contained, established component of the physics of the future. He could not understand how a known professor could so lacking in objectivity, show such "malicious political passion." The room was full to bursting, and from the back rows, it was impossible to see the speaker clearly. Heisenberg was so confused by this pamphlet that he failed to notice that the next lecture was not at all by Einstein but by Max von Laue in his stead. Because of the growing anti-Semitism, Einstein had canceled. The first meeting between Heisenberg and Einstein took place only three years later, in April 1926 in Berlin, where Heisenberg lectured on his quantum mechanics.

Heisenberg in Göttingen and Copenhagen

Heisenberg first traveled from Munich to Göttingen in June 1922 to attend the lectures of Niels Bohr, at the so-called Bohr Festival. Over the course of three weeks, Bohr held a lecture series on the topic "Quantum Theory of the Atom and the Periodic System of the Elements." In one of the lectures, Bohr spoke also about a problem of atomic physics, the "quadratic stark effect," with which the student Heisenberg was familiar. Heisenberg disagreed with Bohr's statement that his assistant, Kramers, had calculated the problem correctly and that the results agreed with the experiments. Bohr was impressed and invited the student for a walk on the Hainberg. They discussed the fundamental physical and philosophical problems of modern atomic theory. Heisenberg observed that Bohr's understanding of the theory was not based on a mathematical analysis but rather on intensive work with actual phenomena. He had achieved his results through intuitive guesswork, not by mathematical derivation.

Heisenberg was very impressed by Bohr's ability "to grasp the relationships intuitively, rather than by formal derivation." Bohr was first of all a philosopher, not a physicist. He gave only positive critiques and gladly acknowledged the work of others. Heisenberg hoped to be invited to Copenhagen soon. He was unaware that the mathematician Richard Courant had written to Bohr that the young Heisenberg was in every respect a really outstanding young man, personally extremely pleasant, not only very imaginative, knowledgeable, and capable but also able to formulate the concepts and to give brilliant lectures on them. He was under consideration as assistant to Bohr. The invitation, however, did not come at once.

During the winter semester of 1922-23 at Göttingen, Heisenberg found time away from his work on his hydrodynamic dissertation to work on the question of how to calculate the energy levels of the helium atom with its two electrons (the simplest atom after the hydrogen atom) in the context of the Bohr/Sommerfeld framework. In his model, however, in contrast to Bohr, he was working with half-integer quantum numbers, which seemed to correspond more closely to the experimental data than the results of Bohr's model, although it was incompatible with the Bohr postulates.

Together with Max Born, then, he tried to describe the quantum states of this atom with two electrons, and to compare it with the observed spectral lines of helium, where calculations had previously not been successful. For this, they employed a mathematical method borrowed from celestial mechanics, the so-called perturbation theory. The result was disappointing. "A comparison [between the theoretical calculations and the experimental data] shows that the result of our investigation is totally negative," write the authors, and "a consistent quantum theoretical computation of the helium problem leads to false values for the energy terms." Thus it was clear that the Bohr/Sommerfeld theory could not stand. Heisenberg returned to Munich to finish his dissertation and wrote to Pauli in March 1923, "Fundamentally, we are both convinced that all previous helium models are just as false as the whole of atomic physics."

However, his work on helium had for Heisenberg the positive consequence that Bohr was so impressed by his efforts that despite the nearly failed doctoral exam in July 1923, in the following winter semester he took him on as his assistant.

The University of Göttingen was a world-renowned center for mathematics and physics. From the time of Carl Friedrich Gauss, viewed as the king of mathematicians (*Princeps Mathematicorum*) in the nineteenth century, this tradition continued through Bernhard Riemann, the inventor of non-Euclidian geometry, which became a prerequisite for Einstein's general relativity theory.

Around 1922, the mathematicians David Hilbert, Richard Courant, and Hermann Weyl worked in Göttingen. Hermann Minkowski, who had found his own path to the relativity theory in 1909, died the same year. David Hilbert was the paramount mathematician of his time and also had a great interest in the application of mathematics to theoretical physics. Heisenberg, who at the time attended Hilbert's mathematical physics seminar, described his feelings about Hilbert's significance in his 1943 obituary:

> *Hilbert's position among physicists and vis à vis physics is doubtless determined by two factors: by the recognition that physics leads to ever new and productive formulations of the problem, that do not arise solely from the fantasy of the mathematician, and by the conviction that the formulations of the problem thus achieved are nonetheless ultimately mastered only through the methods of pure mathematics.*

Hilbert was very interested in the questions of the general theory of relativity and invited Einstein in July 1915 to give six lectures on his theory. He then published his own theory in December that contained the final formulation of the field equations of gravitation. About this, Einstein writes to his friend Zangger: "Only one colleague really understood [my lectures], and he tries to appropriate them skillfully."

The professorships for physics were held by Max Born, James Franck, and Robert Pohl. Max Born had previously occupied himself with optics and relativity theory, but was now interested in the developing Bohr/Sommerfeld quantum theory, and how quantum theory would affect the physics of crystal lattices. When Wolfgang Pauli, who had been Born's assistant after his Munich doctorate, went to Hamburg, Heisenberg was now to continue work on atomic theory. In the winter semester of 1923, he took up the position of Born's assistant at Göttingen, at a time when currency inflation was spreading throughout Germany.

In contrast to Munich, the working atmosphere among the Göttingen group was less open for discussion, less lively. On the other hand, Heisenberg profited from the mathematical expertise at the Göttingen Institute. He wrote home that fundamentally, there were only mathematicians there. After the Bohr/Sommerfeld model of helium, and of the so-called anomalous Zeeman effect collapsed, new ideas were required. Soon after his arrival Heisenberg, together with Born, was able to develop fundamental thoughts on atomic theory. As he wrote in a letter to Pauli, the mechanical conceptions of atoms with electrons in orbits around the nucleus now had only a symbolic meaning. He applied this principle to his special example, the Zeeman effect, where the spectral lines of an atom split in an external magnetic field. That is, one spectral line becomes three.

Fig. 2.6 Heisenberg's Göttingen teachers, Max Born (seated), and James Franck (2nd from right), with visitors Carl Wilhelm Oseen (left), Niels Bohr (2nd from left) and Oskar Klein, scientific assistant to Bohr (right), June 1922, © Werner Heisenberg Estate

In the case of some atoms, a splitting into two or more than three lines was later found. This was called the anomalous Zeeman effect. Heisenberg wrote a long letter to Niels Bohr in Copenhagen about the computations involved in this problem, and Bohr invited him to come to Copenhagen to discuss it. He was able to act on this invitation after the winter semester, in March 1924.

The discussions with Bohr dealt first of all with the fundamental philosophical questions of quantum theory and the definitions of the terms. Next, Bohr addressed a new idea about the emission and absorption of radiation at the atomic level. In classical theory, radiation was a wave phenomenon; in Einstein's explanation of the photoelectric effect, light behaved as though composed of particles – photons. A new visitor to Bohr's Institute, John Slater came up with the idea, as Heisenberg describes in a letter, that there were "waves as well as particles, and the particles were, so to speak, drawn along the waves." Additionally, he introduced a virtual radiation field, whereby atoms were said to communicate among themselves. Bohr and his assistant Kramers took up this idea and drafted a theory of radiation that was named BKS, after its three authors, who were the controversial topic of discussion among quantum physicists until they were disproved experimentally.

On a hiking trip on the "Danish Riviera" in the north of the island Seeland, Bohr communicated a skeptical attitude to his young guest. They were both of the opinion that they remained far from a solution to the problems of quantum mechanics. Bohr himself later noted:

> *Our discussions touched on many problems of physics and philosophy, and particular emphasis was put on the demand for unambiguous definitions of the concepts in question.... We spoke about the fact that here, as in relativity theory, mathematical abstractions might prove useful.*

These ideas would prove exceptionally fruitful during Heisenberg's stay on Helgoland one year later.

First, though, Bohr invited Heisenberg – whom he described as "very brilliant and sympathetic" in a letter to Rutherford – to come to Copenhagen for a longer stay in the fall. Max Born agreed, and so, in September, 1924, Heisenberg arrived back in Copenhagen.

The work on atomic theory, however, initially made no progress. But then a doctoral student of James Franck's, Wilhelm Hanle in Göttingen, discovered a new effect in which the resonance radiation of mercury and sodium atoms in a weak magnetic field was polarized. Bohr explained the effect in terms of classical theory but left it to Heisenberg to carry out the actual computation, who published his results under the title, "On the Application of the

Correspondence Principle to the Question of the Polarization of Fluorescent Light." In Heisenberg's recollection, though Bohr could indeed formulate the essence of the problem with inimitable clarity, he nonetheless shied away from the mathematical abstraction.

In December 1924, Pauli sent his paper on the number of electrons in the various quantum states of the atom to Bohr, and his work was a bombshell: Pauli had assigned to each electron four rather than three quantum numbers, and he postulated that "there can never be two or more equivalent electrons to which the values of all four quantum numbers coincide." The fourth quantum number turned out later to be the intrinsic angular momentum, or spin of the electron. Through this exclusion principle of Pauli's, the number of elements in each group of the periodic system is correctly determined in a simple manner. Pauli did not yet know the reason; that emerged only later from the statistics of systems of multiple electrons. Pauli wrote to Bohr that although his exclusion principle contradicted Bohr's principle of correspondence between classical and quantum physics, it was no greater nonsense than the previous view.

In the final months in Copenhagen, Heisenberg went back to work on "his" Zeeman effect. He successfully compiled the various formalisms in a consistent description. In April 1925, he returned to Germany.

3

The Wonder Years

The Calm Before the Storm of Ideas

Let us pause a moment to consider the situation of these two geniuses at the point at which they were preparing to revolutionize the world of physics. Both were born in medium-sized southern German towns, one of Swabian, the other of Bavarian stamp. Both passed through a Bavarian elementary school, one with Catholic religious training, the other with Evangelical instruction. Both learned to play an instrument and were lifelong practicing musicians, one on the violin, the other at the piano. Both attended a humanistic high school in Munich.

In Heisenberg's high school – 20 years after Einstein attended high school – the curricula had remained essentially the same. Beside the ancient languages Greek and Latin, the only living foreign language taught was French. Both had been especially fond of a teacher of mathematics. Both were fascinated by the laws of Euclidean geometry; Einstein called it the "little holy geometry textbook." For Heisenberg, it was Plato's insight that there were just five regular bodies, and that these could be assembled from triangles and squares. Both were convinced mathematically describable laws underlay natural phenomena. For this reason, both had acquired mathematical and scientific knowledge through extracurricular study, which they found necessary in order to understand natural law.

Whereas Heisenberg's parental home was characterized by humanistic learning, where Greek and Byzantine literature and art were discussed, Einstein had in his Uncle Jakob, by contrast, a scientifically informed interlocutor, who could explain the laws of electrodynamics at the dinner table.

K. Kleinknecht, *Einstein and Heisenberg*, Springer Biographies,
https://doi.org/10.1007/978-3-030-05264-5_3

Albert the student could thus already think through such abstract and absurd ideas as what it would be like to travel on a light beam while observing another light beam. He knew already then that in electrodynamics, a strange number, c, figured prominently, viz., the speed of light.

Einstein's study at the Polytechnic in Zürich was restricted to the framework of classical physics; his diploma work under Professor Weber bored him. In 1905, following several unsatisfactory periods of teaching in Schaffhausen and Winterthur, he was 26 years old, held a good civil service position at the Swiss Patent Office in Bern, and had married Mileva, with a daughter in Novi Sad, and their first son, Hans Albert, at home. For him personally, these were tranquil times, when he could devote himself in his free time to his profound questions and discuss the latest publications in physics with his friends at the Akademie Olympia.

Aside from his two friends, there were no academic people he could talk to. He garnered his knowledge of the problems of modern physics from publications, and in this domain, since his school days and his discussions with Uncle Jakob, he had been fascinated by the generation and propagation of light in space. The foundation of his belief system was the conviction that a divine presence, which he jokingly referred to as "the Old Man," manifests itself in natural phenomena, He did not picture this as a personal god but rather in the Spinozian sense as "the absolutely infinite essence" from which the absolutely valid and immutable geometric propositions and natural laws emanated.

Einstein's discoveries and inventions were known to Heisenberg 20 years later; they were discussed in all the media. Heisenberg's interests went in a different direction; he focused on understanding the structure of the atom. In the summer of 1925, his "wonder year," he was 23 years old and was Max Born's assistant at Göttingen. He had the good fortune of having the most outstanding scientists of his time to converse with, starting with Arnold Sommerfeld in Munich as his doctoral advisor, Max Born as sponsor of his habilitation in Göttingen, Niels Bohr, brimming with ideas, and Wolfgang Pauli as fellow student, an implacable critic, and fount of ideas. After the politically turbulent times that followed defeat in WWI, the outlook under the Weimar Republic promised a hopeful future.

Already at the time of his studies with Sommerfeld his interest was directed towards trying to understand the structure of the atom through analysis of the light emitted by atoms when excited by suitable methods. This area of spectral analysis was the key to understanding the world of the atom. Niels Bohr had proposed a model according to which the negatively charged electrons in the atom revolve in circular orbits around the positively charged nucleus. In this model, only particular orbits were possible, which Bohr specified by so-called "quantum conditions." The model resembled a Keplerian planetary system in which, instead of gravitation, the electrical force was said to provide the attraction between electron and nucleus. Heisenberg's teacher Sommerfeld had significantly expanded and improved Bohr's model of the atom by fitting it to the actual observed spectral lines.

Any physicist could see that the model was incompatible with classical physics, for if electrons move in circular orbits, they are accelerated by the electrical force, and by the laws of electrodynamics would emit X-rays. In the process, they would lose energy and collide with the atomic nucleus. The Bohr atoms are thus unstable. Although the electrons are held artificially in their orbits by the Bohr quantum conditions, these conditions were devised *ad hoc*. Many physicists attempted to improve the Bohr/Sommerfeld model through additional assumptions, but several contradictions to the known laws of physics remained that could not be resolved.

As the two students, Heisenberg and Pauli, hiked along the Bavarian Lake Walchen in 1921, they naturally discussed atomic theory. After a lengthy conversation, the well-warmed hikers came to the conclusion that these Bohr electron orbits could not exist at all, and that any attempt to fix and rescue the model through incremental improvements was futile. A totally new approach was required in order to formulate a consistent theory. But no one knew what such a theory should or could look like. And certainly nothing was changed in Heisenberg's conviction that nature is necessarily so constituted as to be describable mathematically.

Einstein's *Annus Mirabilis*

From the time of his appointment at the Swiss patent office and his marriage to Mileva in 1903, Einstein lived a carefree existence. The office work required of him left time enough to pursue his scientific work. He discussed foundational books on physics and philosophy, as well as the latest publications in physics with his friends Maurice Solovine and Conrad Habicht. Mileva, who ran the household, cared for little Hans Albert. In 1905, two of the three members of the Akademie Olympia were transferred away from Bern, but fortunately a new interlocutor, Einstein's friend Michele Besso, also came to the patent office. With Besso, a mechanical engineer, he could discuss philosophical and scientific problems on their walks home from work together. Besso was his attentive audience, or "sounding board," as Einstein wrote. But although this was not a case of genuine scientific collaboration, Einstein acknowledged him in his paper called the "Electrodynamics of Objects in Motion" for many valuable suggestions. Nevertheless, he missed Conrad Habicht, who was teaching at a gymnasium in the Canton of Graubünden, and complained that he had still not sent him his dissertation. In exchange, he promised him four papers, of which he would soon receive four complimentary copies. These four papers, which Einstein published in 1905, contain things of importance far beyond what anyone could have expected.

The Photoelectric Effect

The first paper "deals with radiation and the energetic properties of light, and is very revolutionary," as Einstein wrote to Habicht. In it, Einstein treats the photoelectric effect, discovered 60 years earlier by the French physicist Becquerel.

Alexandre Edmond Becquerel liked to experiment in the open with sunlight, often in the company of his father, Antoine César Becquerel. In 1839, he noticed a curious phenomenon: a battery yielded more electricity if it was exposed to sunlight. His observation occasioned little interest. Not until 1887 did Heinrich Hertz find that if ultraviolet light impinges on one of two electrically charged metal plates in a vacuum, it induces sparks. After it was discovered that electrons were the carriers of electrical current, J. J. Thomson recognized that in Hertz's experiment, electrons were ejected from the metal by the light. He named the phenomenon the "photoelectric effect," or "photoeffect" for short.

In 1902, in Heidelberg, Philipp Lenard made the crucial observation that electrons emitted from the metal plate under a thousand-fold increase in the intensity of light had exactly the same energy as with lower intensity. "The energy of the electrons displays not the slightest dependence on the intensity," he wrote in the paper. This contradicted wave theory then current. On the other hand, the energy of the electrons was increased if instead of green light, one used blue, violet, or ultraviolet light. That meant the energy of the electrons was proportional to the frequency of the radiated light. In the case of most of the metals used, it turned out additionally that red light was not able to eject electrons, even if the intensity of the light was increased.

In his paper, Einstein at first assumed that the light, in accord with Maxwell's theory (which his Uncle Jakob had already explained to him during his school years, and which since then he had mastered mathematically), was to be understood as a wave phenomenon, where the energy was a continuous function of space. In the paper, he writes that the wave theory of light had proven itself admirably for the description of optical phenomena. It should be kept in mind, though, that the optical observations refer to temporal mean values, not to instantaneous values. One might then take up the idea of Max Planck, who had explained the law of emission of electromagnetic radiation, equally valid for the radiation of heat and of light, by regarding light as a succession of tiny packets of energy, or "quanta."

So if we adopt the heuristic viewpoint, that also in the photoelectric effect individual elementary processes – in which energy is transmitted only in tiny packets – are occurring, we can explain Lenard's observations. Einstein postulated that in the photoeffect, the incident light, which up to that time had been pictured as a wave, consists simultaneously of quanta, or particles. These "photons" are tiny packets of energy whose size depends on the color of the light. They knock the weakly bonded electrons out of the metal. The energy of an electron then is that of the photon, minus the energy of release from the metal. And the energy of a single light quantum is the frequency of the light wave times the number "h," Planck's constant. With his idea, he was able to explain why ultraviolet light can knock the electrons out. The ultraviolet photons have greater energy.

Einstein's heuristic viewpoint was severely criticized. The renowned Berlin physicists around Planck, who would in 1913 propose Einstein for admission to the Prussian Academy, were not persuaded by the light-quantum hypothesis. He had certainly made remarkable contributions to all the important questions of modern physics, but sometimes he missed the mark, as with his light-quantum hypothesis. Planck looked for the meaning of the light-quanta not in the propagation in a vacuum but in the interaction with matter, where light is absorbed and emitted. In propagation of light in a vacuum, Maxwell's laws should be strictly in force.

The experimental physicist Robert Andrews Millikan, who studied the laws of the photoeffect for 10 years in his laboratory, wrote that Einstein's photoelectric equation seemed to describe the experimental results correctly. But the corpuscular theory with which Einstein derived this equation was completely untenable. It contradicted everything we know about the interference of light. For if we send a beam of light through two narrow, closely neighboring slits, we observe a wave-form pattern on a piece of paper set up behind the slits, the interference pattern. This pattern is the same as that produced by overlapping waves of water, and shows that light is a wave. If light consists of tiny corpuscles, however, we cannot explain this interference. It was more than ten years before the experimental results were precise enough for the hypothesis of light quanta to be generally accepted. The reason for the resistance was the aversion of most physicists to the paradox that light could be, in its propagation in a vacuum on the one hand, a wave, but in its impingement on matter, a particle. Indeed, with this hypothesis Einstein created a prime example of the duality of wave and particle, which 20 years later would play a large role in quantum mechanics.

With recognition by the physics community, the path was clear for the awarding of the Nobel Prize to Einstein in 1921, "for the discovery of the laws of the photo-electric effect."

The technical application of the photoelectric effect to the conversion of light energy falling on the surface of Earth into electrical energy was made possible when William Shockley, Walter Brattain, and John Bardeen invented the transistor, made of semi-conductive silicon. Now, the electron released by the photo electrical effect could be used to generate an electrical potential. In 1954, the first solar cell was built from silicon and tested.

The Doctoral Dissertation and the Brownian Motion of Molecules

In June 1905, Einstein submitted his second paper to the University of Zürich, which was accepted as a doctoral dissertation and published in 1906 in *Annalen der Physik*. It dealt with the atomic structure of matter. Einstein made the assumption that the viscosity of water increases if a soluble substance is introduced into it. Using the example of sugar water, from the change in viscosity, he calculated the size of the sugar molecule. In this, the so-called Avogadro or Loschmidt constant N, the number of molecules per mole is employed. The size of the mole for every element and every chemical bond can be derived from the atomic weight. For instance, a mole of hydrogen (H_2) weighs 2 grams, a mole of water (H_2O) 18 grams, and a mole of carbon (C) 12 grams. Einstein's calculations yielded a number of molecules per mole, and thereby the size of the molecule. Apart from a computational error, which he discovered only 5 years later, this yielded the correct scale of the Loschmidt constant.

The third paper, on the Brownian motion of molecules, had a similar theme. If tiny particles are introduced into a liquid, they carry out oscillations ("Zitterbewegung"). These motions can be observed through a microscope, for instance with clubmoss seeds in water. Einstein calculated the magnitude of these oscillations, which increase with rising temperature. He also found that the oscillations increased in proportion to the smallness of the particles in suspension. By mathematical extrapolation, it is possible to calculate how the motion of the invisible molecules in the liquid proceeds, and to estimate their quantity and size. In this paper, Einstein was also able to derive the laws of Brownian molecular motion, which he published in the journal *Annalen der Physik*. Francois Perrin and other researchers then investigated Brownian motion experimentally. The experiments supported the concept (still not universally accepted at the time) of atoms and molecules as real objects, not simply working hypotheses.

The New View of Space and Time

The fourth paper, published in June 1905 under the title "On the Electrodynamics of Bodies in Motion" in the *Annalen der Physik*, has altered our worldview. Electromagnetic waves (thus light, too) that propagate through space at high velocity according to the Maxwell equations posed great problems for the physicists of the nineteenth century. They understood waves in water, and sound waves in air. In these, molecules oscillate periodically at a determined frequency. Now they asked themselves what it was in a light wave that oscillated periodically – what material substance carried out this oscillation. Maxwell himself was convinced that such a substance must exist. He wrote: "There can be no doubt that the interplanetary and interstellar spaces are not empty, but are occupied by a material substance or a body, which is certainly the largest, and probably the most uniform body of which we have any knowledge."

This substance was called ether, a name the ancient Greeks had used to denote the fifth element. This ether was supposed to fill the entire universe, and to be at rest in relation to the fixed stars. In it, the electromagnetic waves were thought to propagate. At the place where the source of light was located, the ether was thought to be perturbed, like an elastic material, and this perturbation was supposed to propagate at the speed of light.

The movement of Earth through this substance would affect the propagation velocity of light. As early as 1881, the American physicist Albert Michelson, working in Hermann von Helmholtz's laboratory in Berlin, tried to measure this effect. In order to rule out perturbation from street traffic, he built his optical apparatus in the astronomical observatory at Potsdam. In his honor, it was later named the Michelson interferometer. The device had two vacuum tubes of equal length arranged in a horizontal plane and perpendicular to each other, with mirrors at the ends into which light rays ran back and forth. In one of the arms of the interferometer, light rays ran in the direction of Earth's movement through space. In the other it ran perpendicular to it.

Were Earth speeding through a stationary ether, the run time of the light in the two arms would have been different. But measurement showed no difference in run times. The same was true with even more precise measurements by Michelson and Morley in 1887. Thus, the hypothesis of a stationary ether was disproved.

As early as 1899 in his study of the physics journals, Einstein had read a paper by Wilhelm Wien that took up the questions of the light ether. On September 28, 1899, Einstein wrote to Mileva from Milan that he had contacted Professor Wien at Aachen about his 1898 paper on the question of light ether. In this paper, ten experiments on light ether with negative results are described, among them the experiment of Michelson and Morley in which the two run times of a light beam in the direction of Earth's motion and perpendicular to it had been measured. Wien writes that using the interferences, the difference in run times would necessarily be observable if light propagated in the ether.

So Einstein knew already in his student years about the result of the Michelson/Morley experiment, and it was also a basis for his reflections in 1905, although he did not refer to the experiment in his paper. In 1950 he still maintained, erroneously, that he first learned of it after 1905. In 1931, he praised Michelson for his wonderful experiments, which had cleared the way for the development of relativity theory by revealing an insidious error in the ether theory. The "wonderful" experiments showed that the speed of light is the same in every environment (physicists say "in every frame of reference"). Whether parallel to the motion of Earth or perpendicular to it, the speed of light always has the same value, 300,000 kilometers per second. According to the previous conceptions of mechanics, Earth's velocity was added to that of light, if the light was emitted in the direction of Earth's motion. What was new in the fact of the unchanging speed of light was that ostensibly the velocities were not added together, in contrast to our earlier conception of space and time.

Einstein was not alone in noticing the contradiction between Michelson's results and the then current ether theory. The Irishman George Francis Fitzgerald had already previously speculated that the only hypothesis that could resolve this contradiction was a change in length of the bodies, dependent of whether they moved parallel or perpendicular to the ether if Earth were flying through a fixed ether. A few years later, the Dutchman Hendrik A. Lorentz came to the same conjecture – that a body of length L oriented parallel to Earth's motion changes its length if rotated by 90 degrees perpendicular to Earth's motion. In this way, the ether theory could be sustained.

Lorentz went further still: he considered two coordinate systems, or "frames of reference," one stationary relative to the ether, and a second moving with Earth. Consideration of two such systems in motion relative to one another has been known since the time of Galileo. If you are sitting in a train at the station and look at a second train on the neighboring track that suddenly begins to move, it can seem that your train has begun to move in the inverse direction. It depends on the frame of reference in which you describe the motion, in your own or in that of the other train. In this description of the motion, the velocities are added: if you walk forward in the train, your speed is added to that of the train; if you walk to the rear, your speed relative to Earth is the difference between the train's speed and your walking speed. The time as you read it on your watch is the same as that on the station clock. These relations between the description in a stationary frame of reference and that in a moving train are called Galilean transformations.

Lorentz found now that if he was to preserve the theory of the ether he had to abandon the Galilean transformations. Lengths in the direction of the relative motion of the two frames of reference shrank; the bodies contracted. He found that now, however, he needed another time measurement in the system in motion. He called the time in the stationary system "general time," and that of a system in motion "local time." Today, the transformations between coordinates in systems at rest and in motion are collectively called the Lorentz transformations. Lorentz could not offer an explanation for these relations on fundamental principles.

Another of Einstein's forerunners engaged with the concept of time was the Frenchman Henri Poincaré. In 1898, he wrote that we have "no intuition about the equivalence of two intervals of time." The simultaneity of two events and the equality of two time intervals must be so defined that the formulation of natural laws are as simple as possible. He was familiar with the Lorentz transformations, and thought them the most satisfying solution at the time. Several years later, he went beyond the concept of local time in that he interpreted it as a physical reality.

He considered two observers, A and B, who move at a constant speed relative to each other and synchronize their watches by light signals. He concludes that all the events in the reference frame of observer B, which observer A measures, run slower relative to B's measurement. But the same is true of the events in reference frame A measured by B. He goes on that because of the relativity principle, the observer cannot determine whether he or she is in motion or at rest. Additionally, Poincaré had the vision to see that one needed to invent a new mechanics, in which the speed of light constituted an insuperable boundary. These ideas anticipate relativity theory. Finally, in 1905,

Poincaré showed mathematically that the succession of two Lorentz transformations is once more a Lorentz transformation, a result Einstein, too, had arrived at several weeks earlier in Bern.

Einstein's Special Theory of Relativity

In June 1905, Einstein sent his paper "On the Electrodynamics of Bodies in Motion" to Paul Drude, editor of the *Annalen der Physik*. In it, he broke radically with the old conceptions of light, space, and time. To begin with, he ejected the nineteenth century ether from physics. Electromagnetic waves propagated in empty space. Second, he postulated that the speed of light was an immutable natural constant. From these two postulates, he deduced that space and time vary if one is sitting in a moving frame of reference. In Einstein's own words:

> *The unsuccessful attempts at verifying a motion of the earth relative to the 'light medium' lead to the conjecture that no properties of phenomena correspond to the concept of absolute stasis (rest), not only in mechanics, but also in electrodynamics.... We wish to raise this conjecture (whose substance in what follows will be called 'the principle of relativity') to a precondition, and in addition introduce the only seemingly incompatible precondition that light in empty space always propagates with a determined velocity V, independent of the state of motion of the emitting body. These two preconditions suffice to achieve a simple, consistent electrodynamics of moving bodies on the basis of Maxwell's theory of stationary bodies.*

To an observer on the ground, a clock in a launched rocket runs slower than an identical clock on the ground. Since at the time rockets did not exist, Einstein illustrated the clock paradox with clocks at the equator and at the North Pole: "A very precise mechanical clock situated at the earth's equator (must) by a tiny amount run slower than one identical, and subject to the same conditions, situated at one of poles."

The concept of simultaneity had to be revised, too. How could the simultaneity of events occurring in different places be determined? The observers had to exchange signals, which at most could be communicated at the speed of light. There is, then, no absolute meaning of simultaneity. An observer in a frame of reference regards two events as simultaneous, but another observer in a reference frame moving relative to it judges it to be non-simultaneous. Lengths in the moving frame contract. But in contrast to the conjectures of Fitzgerald and Lorentz regarding such a contraction, for Einstein it is not the material that contracts through electromagnetic forces but rather space itself which changes in a moving frame. From his principles, Einstein then derives the Lorentz transformations.

The relativity principle, which gave its name to the theory, means that two frames of reference moving uniformly relative to each other are on an equal footing.

We can thus describe the processes of two trains equally well in the reference frame of one train as of the other, or for that matter of the stationary railway terminal. We must note, however, that the time measurements in the two reference frames run differently from one another. In the view of an observer on the platform, the traveler's watch in the train runs slower than the station clock on the platform. This is the famous stretching of time or time dilation of the train in motion.

A puzzle from Einstein's school years led to a further realization. The young Albert had imagined how it would be if he were flying through the universe on a red light beam, and a blue light beam made its way parallel to it. What would he see? Since the other light beam would have the same propagation velocity, it would be stationary relative to the observer, a blue cloud without direction. Since this is impossible, the conception that he himself could move at the speed of light is incorrect. This follows also from the Lorentz transformations. For solid bodies, the speed of light is an insuperable boundary, as Poincaré had conjectured.

Another notable corollary of relativity theory yields the "twin paradox." One twin sets off at high velocity on a lengthy interplanetary voyage, while the other stays behind on Earth. On his return to Earth, the voyager is younger than his brother, who stayed behind. That is, all biological processes in the voyager's body have really run slower than those of his brother. Years later, this paradox was experimentally confirmed using two atomic clocks, one of which circled Earth in an airplane while the other remained stationary.

Space and time are united in the special theory of relativity. Three-dimensional space plus time constitute a four-dimensional space (Minkowski space). The relationship between the space and time coordinates in a stationary reference frame, and one in motion relative to it is described by the Lorentz transformation.

Looking back at this work, Einstein tried in a February 1955 letter to Carl Seelig to contrast his independent contribution to those of his forerunners:

> *Without doubt, the special theory of relativity – when we consider its development in retrospect – was ripe for discovery in the year 1905. Lorentz had already recognized that the transformation later named for him was crucial to an analysis of the*

> *Maxwell equations, and Poincaré had deepened this finding still more. For my part, I was familiar with Lorentz's important work of 1895, but not with Lorentz's later papers, nor with Poincaré's subsequent investigation. In this sense, my work of 1905 was independent. What was new in it was the recognition that the importance of the Lorentz transformation went beyond the context of the Maxwell equations, and related to the nature of space and time in general. Also new was the insight that the 'Lorentz invariance' was a general precondition for any physical theory.*

It is not entirely clear why Einstein here disputes having known Poincaré's papers, even though they were among the reading matter of the Akademie Olympia of the three friends. It may have been because the relationship of the two scholars was not the best. In a report on Einstein for the University of Zürich, Poincaré had once written: "Most of the paths he goes down are dead-ends." Poincaré had never accepted the Einsteinian form of the special theory of relativity. Conversely, despite repeated requests, Einstein had not contributed to the 1919 Festschrift of the *Acta Mathematica* in honor of Poincaré and in an article for *The New York Times* in 1920 had cited only Lorentz and himself as authors of the special theory of relativity.

Mass and Energy

In September 1905, as a supplement to the special theory of relativity, Einstein submitted another paper to the *Annalen der Physik*. Under the title "Is the Inertia of a Body Dependent on its Energy Content?" Einstein drew an additional inference from relativity theory: mass can be converted to energy. The equation that changed the world read: $E=mc^2$. In 1905, Einstein described its significance: "If a body emits energy E in the form of radiation, its mass is diminished by E/c^2. Here it is clearly insignificant that the energy subtracted from the body is directly converted into energy of radiation, so that we are led to the general conclusion: the mass of a body is a measure of its energy contents."

And further, he writes: "An appreciable decrease in the mass of the radium must result. This observation is amusing and intriguing; but whether the Lord God laughed about it, and has been leading me around by the nose is something I cannot know."

The energy content E of a mass m has the enormous magnitude mc^2, where c is the speed of light of 300,000 kilometers per second. If we could convert one gram of hydrogen completely into energy, it would yield as much heat as burning 4,000 tons of brown coal. In accordance with the conservation of the number of nuclear building blocks, however, the laws of nature do not admit of this process.

That a similar process called nuclear fusion is possible, and which takes place in our Sun and in all the stars, though, was recognized by Hans Albrecht Bethe and Carl-Friedrich von Weizsäcker in 1937. In the nuclear fusion process in stars, the element helium is created from hydrogen at temperatures of 15 million degrees, and heat and radiation energy are released. The difference in the masses from beginning to end state in this process is, according to Einstein's famous equation $E=mc^2$, converted to energy. The nuclear fusion processes in the Sun are the basis of the Sun's energy. Absent this phenomenon, i.e., without the conversion of mass into energy, our life on Earth would not be possible. The relevance of relativity theory for our life on Earth is thus apparent.

A second way to convert mass into energy was discovered by Otto Hahn and Fritz Straßmann in 1938. They found that the heavy nucleus of the element uranium splits into two parts if it is bombarded with slow neutrons. This is called "fission" of uranium. In this case, too, the particles created are together somewhat lighter than the original material, and the lost mass is converted to heat. Einstein had anticipated such a process when he spoke of an appreciable decrease in mass in the case of radium. The energy released in the form of the kinetic energy of the fragments and the fission neutrons present is on the order of 50 million times greater than that of the chemical combustion of a carbon atom. Einstein said of this, "The discovery of nuclear fire is the greatest invention of mankind since the practical application of fire."

Reactions to the Revolutionary Papers

It is remarkable that Paul Drude, editor of the leading physics journal of his time, accepted all the papers for publication. Einstein was after all still unknown in the scientific world; he himself was not sure such revolutionary papers would be accepted by the editors. One of the editors, Max Planck, had read them at the same time. Amazingly, Planck had reservations about the light quanta hypothesis, even though it had been inspired by his own work.

On the other hand, Max Planck was the first to recognize the importance of relativity theory, and he helped find acceptance for it against widespread skepticism. As early as March 1906, just half a year after the appearance of Einstein's revolutionary work, Max Planck delivered to a conference of the German Physical Society a lecture on relativity theory that in the opinion of Max von Laue was "unforgettable to all participants", and challenged the audience to test its consequences experimentally. In the spring of 1906, he wrote Einstein a letter in which he praised his work in glowing terms. This was the beginning of a collegial friendship that would last for decades.

In a lecture given in the United States, Max Planck compared relativity theory to the Copernican revolution. Just as Copernicus had liberated us from the conception that Earth was in stasis (at rest), after Einstein we had to abandon the idea of absolute time and absolute space.

Professor in Zürich, Prague, and Again in Zürich

After the explosion of his creativity in the wonder year 1905, Einstein took his doctorate in 1906 with a dissertation on the magnitude of the atom, which was submitted to the University of Zürich. Although he did not complain about his work as "distinguished, properly remunerated, Swiss ink-shitter," a situation which, incidentally, gave him the opportunity of riding his physics hobbyhorse, and "sawing away on the violin," his goal was an appointment as a professor at a university. He wanted to return to the academic world. As an interim step, he considered seeking a teaching position and asked his friend Marcel Grossmann how that might be done. An application to the Canton school in Zürich was unsuccessful, however.

Another path to his goal was to meet the qualifications for becoming professor, the "habilitation". He had become known through his papers, and Professor Alfred Kleiner at the University of Zürich advised him to apply to the University of Bern to gain experience in lecturing. So he submitted his application for habilitation there, to which he attached his doctoral dissertation and 17 publications, among them the trailblazing papers from 1905. After 4 months' consideration, the faculty of natural sciences of the University of Bern declined the application because the formal requirement of an independent habilitation publication had not been fulfilled. So Einstein was compelled to submit another, yet unpublished paper to finally become an independent lecturer. He was then allowed to deliver lectures without an appointment at the university. He took these duties on himself alongside his work at the patent office in order to qualify for a professorship. At first, he delivered his lecture to an audience of three at 7:00 in the morning before going to work. He would later switch the time to the evenings.

In 1909, an associate professorship in theoretical physics became available at the University of Zürich. Two candidates were on the short list, Friedrich Adler, an independent lecturer at the University of Zürich, and Einstein. The majority of the members of the school board belonged to the Social Democratic

Party, as did Adler. He was an adherent of the positivism of Ernst Mach, as well as of the dialectical materialism of Marx and Engels. He had written a polemic against Vladimir Ilyich Ulyanov, called Lenin (who was living at that time in Geneva, Switzerland), who had condemned Ernst Mach's philosophy as reactionary and incompatible with dialectical materialism. When Adler learned he was placed first on the list, he protested with the words: "If a man like Einstein is to be had for our University, it would be folly to appoint me. My qualifications as a researcher in physics are not remotely comparable to Einstein's".

Thereupon Einstein was appointed.

Fig. 3.1 Albert and Mileva Einstein in Zürich, 1910

Initially, a smaller salary was offered him than at the patent office, so it was a struggle to maintain his previous level. On the other hand, he was naturally happy to be rid of his daily eight hours at his desk at the patent office and aside from his regular lectures to be able to devote himself to his own research interests. To be sure, he also had to participate in the activities of the faculty. There were administrative duties for which he cared little, because they cost him time and kept him from his research work. As compensation for this, he was able to live in Zürich again.

His new position did require him to adapt his Bohemian life style more in keeping with the bourgeois habits of a Zürich professor. That was not entirely successful. The Einstein family moved into the same house in which Friedrich Adler, his former rival for the professorship, lived with his family. The Einsteins got along well with them; Adler thought that "their Bohemian lifestyle is similar to ours."

A short time later, Adler became editor-in-chief of a Social Democratic newspaper and gave up science. He returned to Austria and became the Party secretary. For political motives, he murdered the Austrian Prime Minister Karl von Stürgkh. Einstein offered to serve as character witness for Adler. Initially, Adler was sentenced to death. Then the sentence was commuted to imprisonment, and he was ultimately released in 1918.

Einstein's lectures in Zürich were better attended than in Bern, and his audience was comprised of "real" students. Preparation for such lectures demanded a great deal of work, especially when giving them for the first time, and this was surely true for Einstein, too. He made an effort to be stimulating for his students. He succeeded, and when that summer it was rumored that Einstein had received an offer from Prague, the students wrote the Department of Education asking that everything necessary be done to retain this outstanding research scientist and lecturer. However, when the offer had actually come, Einstein accepted the call without giving those in Zürich the opportunity to enter into negotiations for him to stay. The fact that he was made full professor and a voting member of the faculty in Prague may have figured in this.

Prague was the oldest university in central Europe, founded in 1348 as Universitas Carolina by Kaiser Karl IV, with lectures given in Latin. In the eighteenth century, German became the principal language of instruction, even though the majority of the population spoke Czech. From 1882 on, the university was divided into a German and a Czech part. The first rector of the German university was Ernst Mach, whose papers Einstein had read. In the fall of 1910, the position of full professor for theoretical physics fell vacant. The physicist Anton Lampa, an ardent adherent of Mach's philosophy, saw an opportunity to appoint a physicist sympathetic to Mach's ideas, and he persuaded the faculty to appoint Einstein. In Austria-Hungary, the final word, however, lay in the hands of Kaiser Franz Josef in Vienna. It was his view that only a person who belonged to a recognized religious community could be appointed professor. Although since his flight from Munich, Einstein was technically without a religion, he was persuaded now to declare himself a member of the "Mosaic faith."

After assuming his post, Einstein took no part in the running disputes between the German and Czech divisions of the university. He observed the conflicts more as an amused onlooker. A large segment of the German professors were of Jewish heritage. The assistant whom Einstein engaged also came from a Jewish agricultural family. Einstein learned from him that in the village communities, the Jewish farmers and merchants spoke Czech in daily life, but on the Sabbath, only German.

The Jews of Prague were a respected minority who, together with the German minority, stood out from the Czech majority, and had little contact with them. Einstein became acquainted for the first time with a Zionist group around the librarian Hugo Bergmann, to which the writers Franz Kafka and Max Brod also belonged. They sought to create their own Jewish cultural life in art, literature, and philosophy – liberal, not orthodox. It was to be open also to other, especially German, philosophical currents. Bergmann tried to win over Einstein for Zionism, but he was not yet interested.

The meeting with Einstein moved Max Brod to model the character of Johannes Kepler in his novel *Tycho Brahes's Path to God* after Einstein. The juxtaposition of the two astronomers embodies the conflict between the experienced elder, who is not quite prepared to surrender the old world view, and the young, creative spirit, who succeeds in solving the puzzle of planetary motion by abandoning the conception of Earth as the center of the universe. Tycho envied Kepler for the natural, respectable, humane manner in which he had gained fame. It is true, Brod characterized him also as insensitive and cold. He has Tycho say: "You take nothing into consideration, take your own sanctimonious path…. You do not actually serve truth, but just yourself such is your purity and your integrity."

On the question of whether we should preserve the old, geocentric system for the sake of the princes and the Bible, thinkers were divided, and Kepler does not show such considerations. His (and Brod's) affirmation of the Copernican turn, the revolution of the geocentric world view is expressed thus: "We have to please only the truth, and none other."

Einstein himself, however, was already moving toward an elaboration of his theory. While the world was just becoming accustomed to living with it, he recognized its weaknesses. The theory dealt only with frames of reference in uniform motion, independent of forces. But what happens under the influence of forces – gravity, for example? He pondered whether a light beam is deflected from its course under the influence of gravity, and found that in accord with his equivalence principle, the light beam ought to be deflected in the direction of gravity, and that it would be possible to observe the phenomenon in fixed stars near the Sun during a solar eclipse.

Einstein's residence in Prague lasted only a year. To be sure, he had a splendid institute and a spacious study with a view onto a large park with people strolling by. When the Austrian physicist Philipp Frank visited him there, he remarked humorously, "There you see all those crazy people who don't concern themselves with quantum physics." It was the garden of a psychiatric hospital they were looking down upon. Despite the comfortable circumstances at the Institute, Einstein's assessment of life there was not quite as enjoyable as life in Switzerland. He wrote to Lucien Gavan: "Most of the population does not speak German, and they are hostile towards Germans. The students are less intelligent and ambitious than in Switzerland, too."

The conventions of formal behavior in the Austro-Hungarian monarchy were an anathema to Einstein. He hated "the snobbery and pretention, and the caste system." So he wished to return to Zürich. In the meantime, the Swiss Polytechnic had been rechristened the Swiss Technical University (ETH), in other words, a proper university with the authority to award doctorates, and offering the best working conditions, since, unlike the University of Zürich, it was not financed by a Canton but by the nation at large.

From Prague, Einstein tried to get an appointment at the ETH together with his friends Marcel Grossmann and Heinrich Zangger. Zangger wrote to the Federal President Ludwig Forrer, attestations were gathered from all over the globe, and a new professorship in theoretical physics was created especially for Einstein. In January 1912, he was appointed the first full professor of theoretical physics at the ETH. He was not obliged to lecture to beginning students, nor supervise laboratory experiments. He had only to hold occasional seminars or lectures for a few advanced students. He had optimal working conditions and a top-notch salary. Nonetheless, his stay at the ETH lasted only a little over a year also. The luminaries of physics wished to recruit him for Berlin. Max Planck and Walter Nernst traveled to Zürich to tender Einstein an offer.

This was made possible through a new society, established on the initiative of Kaiser Wilhelm II, the Kaiser-Wilhelm-Society (KWG). Outstanding scholars were to be attracted as members of the society, and absent lecturing duties allowed to devote full time to their research. It was prestigious for bankers, industrialists, and businessmen to be members of this society, and to contribute large sums for scientific research.

The mission of the two envoys was to win Einstein for Berlin. Although there was as yet no institute for physics, Einstein was to become director of the Kaiser-Wilhelm-Institute for Physics to be created, and simultaneously a member of the Prussian Academy of Science and Professor without instructional duties at the University of Berlin, established by Wilhelm von Humboldt. An initial salary would come from the academy, and a second from the banker Leopold Koppel.

The two Berliners vividly described to Einstein the advantages of a job in their city. He could devote himself totally to his research, free of instructional duties, but would find stimulation for his work in discussions with the many leading physicists, mathematicians, and chemists there.

Einstein asked for 24 hours to think it over, which the envoys took advantage of for an outing up Mount Rigi, with its view over Lake Lucerne. When the train pulled into the Zürich railway station that evening, Einstein waved a white handkerchief, the pre-arranged signal of his agreement to go to Berlin.

The General Theory of Relativity and Berlin

Einstein arrived in Berlin early in 1914. What motivated him to exchange his princely position in his beloved Zürich for an equally excellent position in Berlin? Since his time in Aarau, he had learned to value the advantages of democratic Switzerland over the German Reich. Now, suddenly, Berlin was more attractive, so attractive that he did not even give the Swiss authorities the opportunity for negotiations to persuade him to stay.

One reason was surely the great importance that science, and especially physics, had in Berlin. In a letter to his cousin Elsa, Albert stressed the "colossal honor" that this appointment meant for him, and the freedom the completely independent position at the academy offered. The international fame of Max Planck, the inventor of quantum theory, the great number of famous researchers at the University of Berlin, the esteem enjoyed by scientists among the general public in the imperial capital – all this made Berlin a mecca for research in the natural sciences. To this was added the tailor-made working conditions with no instructional duties, no administrative duties, but full faculty rights. In fact, the domicile of the "Kaiser-Wilhelm Institut für Physik" was Einstein's apartment until much later, when a building was constructed. Einstein saw the chance here to devote himself predominantly to the completion of his difficult work on the general theory of relativity.

That would probably have been enough to account for the move. A further decisive point was the deteriorating relationship with Mileva, from whom he wanted to separate, and a new love for his cousin Elsa, with whom he had been friends since childhood. Her father, Rudolf Einstein, was a cousin of Albert's father, Hermann. Her forebears came from the same extended Einstein family in Buchau as Albert. She had separated from her husband, Max Löwenthal, in 1908, and lived with her two daughters in Berlin in her parents' house. Einstein had seen her again already in April, 1912 during his Prague residence during a visit in Berlin, and had immediately fallen in love with her. At the time, he wrote from Prague that he had become so fond of her in just those few days that he could scarcely find words to express it.

In March 1914, Einstein arrived in Berlin alone; Mileva and the children followed in May. The atmosphere in the apartment they moved into in Dahlem soon became so uncomfortable that first he, and then Mileva and the children, moved out and found lodging with friends and relatives. At this point, the marriage existed only in name. In the summer of 1914 Einstein decided to divorce Mileva. Their friend Michele Besso traveled from Zürich to fetch Mileva and the children; Einstein saw them off at the railway station. The following day, he wrote to Elsa that the final battle had been fought; during their farewell at the station he had wept bitterly, but he knew it was the best path for him, even though it would alienate the children. He stressed that now Elsa had proof that he could make sacrifices for her.

After Mileva's departure, Einstein felt freed to resume work on the general theory of relativity. While opinions about the special theory of relativity throughout the world differed since it was published in 1905, it was clear to him that it was incomplete. For it made statements only about a world absent forces, in which objects moved in straight lines and uniformly. He began to ponder how things would change if a force operated on the bodies. For him gravity, whose effects acted on Earth and in the universe was the most important force.

He had been working on the formulation of a theory of gravity for seven years. His "lucky moment" came in November 1907, just two years after the Wonder Year of 1905, as he sat at his desk in the patent office in Bern. He wrote later, "Suddenly, I had an insight. If a person is in free fall, he will not feel his own weight. I was taken aback. This simple thought experiment made a profound impression on me. It led me to a theory of gravitation." From this thought experiment, Einstein derived a principle on whose foundation he was able to build the general theory of relativity: the equivalence principle.

We can clearly understand the principle if we pursue Einstein's "insight" further. In a first experiment, there is an elevator of cosmic dimensions whose cable is severed, and it is in free-fall. In the elevator are physicists who are carrying out their experiments in total calm. They take pens, coins, and keys out of their pockets and let go of them. But nothing happens: pens, coins, and keys remain floating in the air because, in accordance with the Newtonian law of gravity, they are falling at the same speed as the elevator. The physicists might think they have escaped Earth's gravity and are floating in empty space. In fact, they cannot tell whether they are in a falling elevator or in gravity-free space.

In the second thought experiment, the physicists find themselves in a rocket in space that is thrust upwards at a constant velocity by its engines. They feel a downward force, and all objects fall to the floor. The physicists cannot tell whether they are standing in a rocket set on the ground or in a rocket traveling at a constant acceleration in free space. A downward force is acting on all objects, either gravity or the inertial force caused by the rocket's motion.

The equivalence principle derived from this states that the weight of the mass subject to gravitation, and the inertial mass of a body, are equivalent and indistinguishable. From this principle alone, several consequences are derivable, independent of a mathematical theory.

One is the deflection of light in a gravity field. If a light beam is sent horizontally from one wall to the opposite wall in a rocket accelerated upward, then when viewed from outside, the light beam will bend downwards, because during the transit time from one wall to the other the rocket has accelerated upwards. Because of the equivalence of acceleration and gravitation, this means that the light beam in the gravitational field of a mass will also be deflected toward it. This would be observable during a solar eclipse in stars at the edge of the Sun.

He could deduce a second consequence directly from the equivalence principle: a small slowing of the running of clocks located in a strong gravitational field. The equivalence principle states that the gravitational force is equivalent to the acceleration of a rocket, and in an accelerated rocket, clocks run slower. If the clocks run slower, however, the number of oscillations per second is shifted, i.e., the frequency of a light wave. The spectral line of an atom located in a strong gravitational field changes its frequency, and thereby its color, from violet to blue, or blue to green, or green to yellow and red. This gravitational red shift in spectral lines emitted from the surface of the Sun ought to be measurable. But it is much smaller than the classic red shift of light from stars moving away from Earth.

Because of the diverse duties in lecturing and faculty meetings during his tenure as a professor in Prague, Einstein was unable to devote himself to the elaboration of relativity theory. Still, he could discuss with his friend, the mathematician Georg Pick, which mathematical methods this problem might be approached with. Pick was well versed in the extension of Euclidian geometry, through the differential geometry of the mathematicians Bernhard Riemann, Gregorio Ricci-Curbastro, and Tullio Levi-Cività. In this mathematical discipline, space can be curved; the angular sum of a triangle is no longer 180 degrees. Einstein had to take up this mathematics.

Moreover, he found time to calculate the most important consequence of the equivalence principle, the deflection of starlight in the gravitational field of the Sun, and to publish his results in the *Annalen der Physik*. At a total solar eclipse, the positions of the stars visible at the edge of the Sun ought to shift outwards to an observer on Earth. Einstein calculated the deflection of light from such a star. He published the value he found of the angular change as 0.87 arc seconds, although shortly before publication he had discovered that his calculation still rested on Newtonian theory supplemented by the equivalence principle. That did not bother him. His maxim was, "He who has never made a mistake, has never tried something new."

Einstein thought the astronomers could work on this question and clarify it experimentally, even if these considerations might appear "adventuresome." Later, after the general theory of relativity was completed, Einstein determined that this value was incorrect, and in actuality it had to be twice as large.

After his return to Zürich from Prague in 1912, he turned to the mathematical elaboration of his theory of gravitation. As we now know from his scientific diary, for his gravitational theory he tried to utilize the mathematics of curved spaces, which he had heard about in Prague. This turned out to be extremely difficult because he did not command the mathematical equipment for it. In his plight, then, he turned to his friend Marcel Grossmann, who had become a professor of mathematics at ETH, and who introduced him to the tensor calculus of differential geometry, which is an extension of the more familiar Euclidian geometry which we know from school.

On a Euclidian surface – a sheet of paper – two parallel lines meet at "infinity." The sum of the angles of a triangle is 180 degrees, and the distance between two points is the length of a straight line between the two points. None of this remains in effect if the paper is curved and, say, takes the form of a spherical surface. Then, the shortest connecting line between two points is not a straight but a geodesic line. Correspondingly, a three-dimensional space can be curved, as can a four-dimensional Minkowski space, in which the special theory of relativity operates. Of course this is difficult to picture and can be described exactly only mathematically. Distances between two points in such spaces are determined by a so-called metric, and it was Einstein's idea that this metric is determined by the masses situated in the space.

It emerges from the Zürich diary that Einstein had discovered the proper form of the field equations of gravitation as early as 1912, but that he then later discarded them. He began the search for the "correct" equations from the beginning again, and fell into a burst of creative energy. He worked his way through the thickets of tensor calculus, and rejected all diversions such as invitations to lecture. Never in his life had he so tormented himself as with this theory of gravitation. Compared to it, the special theory of relativity was child's play, as he wrote Max Planck. It would take another three years of hard work, with detours down many false paths, before he reached the final form of the theory.

The work was interrupted by the move to Berlin and the divorce from Mileva. Thereafter, the way was clear again for uninterrupted, intensive work for Albert the bachelor. He did not spare himself; he worked like a horse, smoked like a chimney, ate indiscriminately, and slept irregularly, as he reported to Elsa. After another year, he spied land: he had found a definitive form of the field equations.

On November 25, 1915, Einstein lectured on his results for the field equations of gravitation to the Prussian Academy, closing triumphant and relieved with the sentence: "Herewith, finally, is the general theory of relativity completed as a logical structure."

The theory offered a new conception of space and time. Space, Einstein concluded, has a geometry that is determined by the masses within it. It is the masses that cause the curvature of space. Einstein repeated the experimental consequences following from the theory. He could explain the deflection from Kepler's elliptical form of the orbit of the planet Mercury. The orbit is not a closed ellipse. Instead, the point of the orbit closest to the Sun, the perihelion, wanders in the course of a century by a tiny angular amount around the Sun. The greater portion of this anomaly can be explained by the influence of the other planets. There is a remainder, however. Whereas it is inexplicable in Newton-Kepler mechanics, Einstein was able to calculate that this remainder of the rotation of the perihelion is accounted for exactly by the general theory of relativity.

The second spectacular conclusion from the theory was the deflection of light in the gravity field of the Sun, which Einstein had calculated earlier. The tracks of light near the Sun are not straight lines but are curved and are bent towards the Sun. According to the final theory, the angular deviation was twice as large as in the earlier paper, and ought to be observable at a solar eclipse. Such an expedition was not to be thought of during the war, but it was known that in March 1919 a total eclipse would offer particularly favorable conditions for observation because especially bright stars would be near the Sun.

Three weeks after the armistice in November 1918, Arthur Eddington announced that the British Astronomical Association was planning two expeditions to observe light deflection near the Sun. Eddington was a Quaker, a religious community that even during the war made an effort not to demonstrate hate towards Germany. To him, science was a bridge of understanding among peoples, and the confirmation of the theory of a German scholar seemed perfectly proper. So the BAA in London equipped two expeditions to those positions at which the total solar eclipse could be observed.

One expedition went to Principe Island in the Gulf of Guinea, a Portuguese colony; the other went to Sobral in northern Brazil. Eddington himself joined the expedition to Principe. Even though the weather on this day was unfavorable, and at the start of the eclipse the corona of the Sun was partially hidden behind clouds, the astronomers were able to use the 300 seconds of total darkness to expose 16 photographic plates. Important stars were obscured by clouds on many of the plates, but five bright stars were visible on one plate sufficient to measure the deflection. The astronomers of the Brazil expedition also brought home usable photographic plates.

After their return to London, the astronomers compared the photographs of these stars from previous observations in the night skies with those pictures captured at the solar eclipse. It was clear that the stars on the photographic plates of the eclipse were further from the Sun than those on the sunless night photos. The deflections in the two observations measured 1.6 arc seconds and 1.98 arc seconds, respectively, with a margin of error of 0.3 arc seconds. The results corresponded pretty exactly to the value predicted by Einstein of 1.7 arc-seconds. The Royal Society and the Royal Astronomical Society announced a joint meeting to take place on November 6, 1919, at which the results of both expeditions to observe the solar eclipse would be announced.

At the meeting, the Astronomer Royal announced that the observations of the expeditions confirmed Einstein's theory of gravitation. At this traditional ceremony, in front of a portrait of Isaac Newton, the scientific community – and soon thereafter, the world – learned that after 200 years, Newton's mechanics would have to be revised on account of a theory developed, of all things, in hated Germany. The president of the Royal Society, Sir J. J. Thompson, opened the meeting with the words, "Einstein's teaching is one of the greatest achievements of human thought." In the course of the meeting, however, he admitted that he himself had not grasped the sense of Einstein's theory.

The next day, *The London Times* ran with the headline: "Revolution in Science. New Theory of the Universe. Newton's Ideas Collapsed." *The New York Times* also put the news on the front page, and the German newspapers followed some days later. On the front page of the *Berliner Illustrirte Zeitung* in December, the text under a photo of Einstein read: "A new dimension of world history: Albert Einstein, whose researches mean a total upheaval of our view of nature, and are of equal value to those of Copernicus, Kepler, and Newton."

Isaac Newton was toppled from his throne; Einstein had finally become a world famous figure. Every utterance from his mouth was eagerly snatched up by the press, and on every imaginable topic, too.

The great publicity of the relativity theory rested also on the fact that, after four years of war and revolution in Russia and Germany, people were profoundly unsettled, and now had the impression that in science, too, the old order had been shaken. On the basis of a misunderstanding, journalists came up with the slogan "Everything is relative," which was further interpreted to mean "revaluation of all values." The political situation in 1919 fit such far-reaching interpretations.

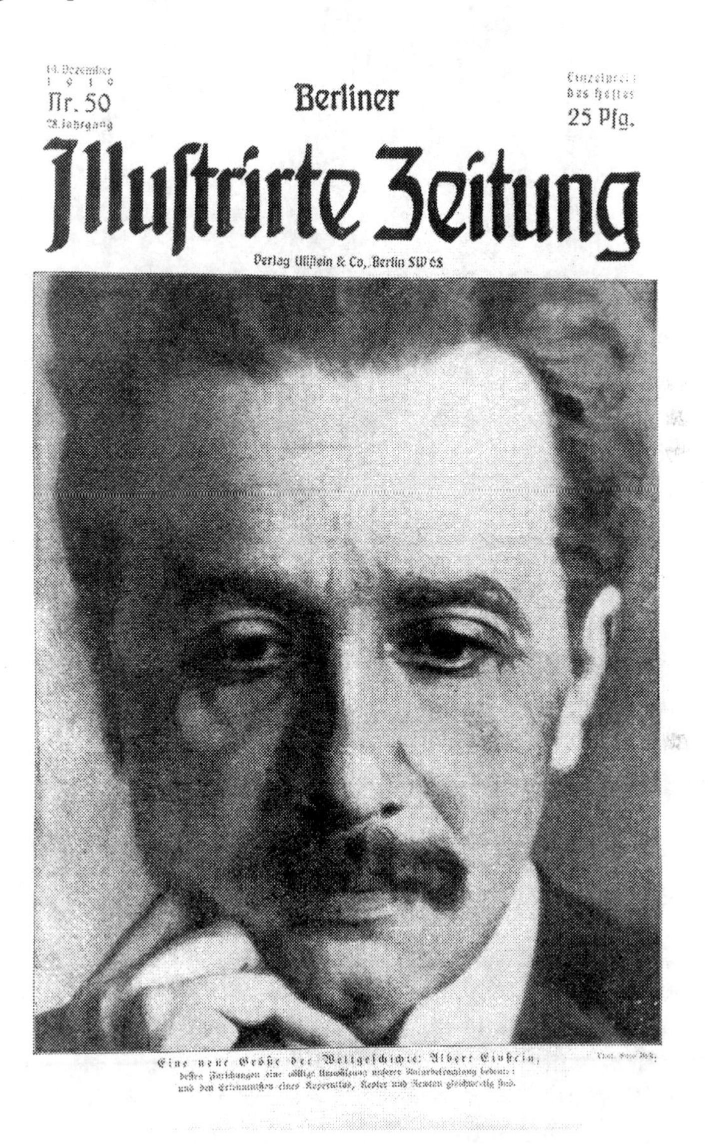

Fig. 3.2 *Berliner Illustrirte*, 1919: A great new figure of world history

Fig. 3.3 Einstein in the library of the Kaiser Wilhelm Institute for physics, 1921

The general theory of relativity, which had been brilliantly confirmed by the experiments with light deflection, rested on what Einstein had named the equivalence principles. Many further consequences of the theory were at this time still not demonstrable. Only in the course of the century did the whole scope of the edifice of ideas emerge.

Consequences of the General Theory of Relativity

The consequences were far reaching. It was now possible to describe the evolution of the universe, and it emerged that it was expanding. This finding resulted from the red shift in the spectral lines emitted by stars discovered by the astronomers Milton Humason and Edwin Hubble in 1927 and 1929, respectively. The further away they were, the faster they moved away from us, and the more the light emitted by them was shifted to the red end of the spectrum. Our position was no longer relevant; rather, every position in the finite but unbounded universe was on equal footing. Each galaxy was moving away from every other one at a velocity proportional to its distance, like two points on the surface of an expanding balloon. If this was the case, the galaxies must originally have been closer together. From the Hubble constant, we can calculate that this was true approximately 13.8 billion years ago. This was the time of the "Big Bang."

Furthermore, it followed from the general theory of relativity that a very large mass can capture light permanently; in other words there can be "black holes." These were subsequently found, in fact. Such a black hole is located at the center of our Milky Way, with a mass of 3.8 million suns, around which the stars in its vicinity circle, as Reinhard Genzel of the Max Planck Institute for extraterrestrial physics observed using an infrared telescope.

Clear evidence of the Big Bang was furnished by the discovery of the cosmic background radiation by Penzias and Wilson in 1965. This microwave radiation strikes Earth from every direction. The temperature of this radiation corresponds to 2.7 degrees above absolute zero, or about minus 270 degrees Celsius. We regard it as the cooled light particles from a distant prehistory, only about 380,000 years after the Big Bang. They reveal to us how the formation of chunks of matter occurred. The temperature of this radiation displays tiny fluctuations of around 1/100,000 degrees. From this we deduce how much matter of the kind we are familiar with was present at that time. It constitutes just 4 percent of the total mass of the universe. The rest, whose nature we do not yet understand, consists of non-luminous, "dark" matter, and – as has emerged in recent years – of mysterious dark energy.

This knowledge about the formation of the universe has totally transformed our world view. Philosophy and theology also treat this question. At the same time, relativity theory has had more practical effects; for example, short-lived particles from cosmic radiation created at an altitude of 10 kilometers at the edge of the atmosphere arrive at Earth's surface through time expansion, even though they ought to disintegrate after a much shorter flight path. Machines can be built that accelerate particles to near the speed of light. The satellite-aided GPS system could not function without relativity theory. Mass is converted to heat and electricity in nuclear power plants.

Einstein himself once summarized the general theory of relativity thus: "Previously it was thought that if all objects were removed from the Christmas [tree] room, we were left with an empty room. Now we know that when we remove all masses, there is no longer any space."

He presumed there was a plan underlying the laws of nature, and he wanted to discover it. He "wanted to understand how the Old Man had made the world."

Heisenberg's Breakthrough to Quantum Mechanics

While Einstein concerned himself with the largest structures and put the physics of the cosmos on a new foundation, Werner Heisenberg was interested in the smallest building blocks of matter, atoms. From his years of study with Sommerfeld, he was convinced that the Bohr model of electrons orbiting around the atomic nucleus could not be correct. Nor did the elaboration of the atomic model by Sommerfeld, in which he himself had participated as a student, alter in any way the fundamental difficulty, viz., that elementary principles of physics were violated by this approach. The orbits were not stable by themselves but were stabilized only because Bohr complemented Newtonian mechanics with several arbitrary quantum conditions. Accordingly, each electron was said to adopt its own Bohr orbit, characterized by quantum numbers.

Even if one accepted the quantum conditions as a postulate, there remained many difficulties in the interpretation of spectral lines emitted by the atoms. There were several phenomena Heisenberg could explain only if he used half integers as quantum numbers in characterizing the orbits instead of whole integers, as postulated by Bohr.

When the behavior of atoms in a magnetic field was examined, new difficulties emerged. The spectral lines observed in a spectrograph change; the lines split up, as was first observed by the Dutch physicist Zeeman. The interpretation of the so-called Zeeman effect occupied Heisenberg for a long time, and problems arose in the process. He named this issue his "Zeeman salad."

Still greater difficulties manifested themselves dealing not just with a single electron orbiting its nucleus, as in the case of hydrogen, but with two electrons, as in the helium atom. How would these two electrons interact? Paul Dirac, who later would successfully establish the connection between relativity theory and quantum mechanics, later wrote:

> *At that time, the young people tried to describe the helium atom by setting up a theory of the interaction of the Bohr orbits, and doubtless would have pursued this direction had not Heisenberg and Schrödinger come on the scene. They would simply have reckoned for decades on the interacting Bohr orbits, and many people would have taken up this task, and, independent of one another, have adjusted their assumptions to match the results of their calculations with the latest experimental data.*

Conservation of Energy and the Compton Effect

Even the doyen of atomic theoreticians, Niels Bohr in Copenhagen, whom all young physicists revered as an authority, did not know how to proceed. In a lecture in February 1925, he said: "In our attempts to develop an atomistic interpretation of the directly observable phenomena…. We have to relinquish the conceptions on which the previous description of natural phenomena rested. Our current terms do not permit a description of atomic processes that conform to the law of conservation of energy, which holds so central a role in classical physics."

He pondered whether this sacrosanct principle had to be partially abandoned. Bohr speculated that the law was valid only for statistical mean values but in individual elementary processes could be violated, and together with his assistants Kramers and Slater issued a corresponding theoretical paper.

One such elementary process was the scattering of light in the form of X-rays on electrons, the so-called Compton effect. In the process, light changes its wavelength. If Einstein's hypothesis of the quantum nature of light, that is, the existence of energy packets of light in the form of photons, was correct, and if the energy and the momentum in every elementary process was conserved, that could be tested experimentally. Walther Bothe and Hans Geiger in Berlin had the idea of detecting both the scattered photon and the triggered electron, and determining whether these two reaction partners appeared simultaneously. They invented the "coincidence method," which won Bothe the Nobel Prize in 1954. The experiment showed that energy and momentum were conserved in the individual elementary process, and that Bohr's resort was a false path. Nonetheless, he continued to maintain that "the corpuscular nature of radiation lacked a sufficient basis." However, he withdrew a paper on this subject.

Wolfgang Pauli in Hamburg was pleased about the confirmation of the quantum nature of light and the conservation of energy in the Compton effect, and that thus the Bohr-Kramers-Slater theory would no longer be able to impede progress in physics. But he himself did not know a way out of the impasse. In May 1925, Pauli wrote to Bohr's assistant Ralph Kronig in Copenhagen. "Physics at the moment is once again very muddled. In any case it is much too difficult, and I wish I were a film comic or any other thing and had never heard of physics. Now I hope, though, that Bohr rescues us with a new idea. I make this fervent appeal to him,"

But salvation would come not from Copenhagen, but from Göttingen and Helgoland.

Burst of Creativity on Helgoland

In the spring of 1925, the 23-year-old Heisenberg, assistant to Born and already an independent lecturer at Göttingen, suffered from a persistent hay fever. In order to distance himself as far as possible from blooming shrubs, he traveled in June to the North Sea island of Helgoland. There he fell into a veritable frenzy of creativity. In his memoires, he compares this to one of his treks in the mountains. As they climbed upwards, the fog grew ever thicker; the group came into a completely confusing jumble of boulders and mountain pines, in which they could no longer pick out any path. Nonetheless, they climbed on. The fog grew thicker, but from above more light came into view. And after a further steep climb, they stood on the summit of an outcropping above the sea of fog in the "sunlight of understanding."

The basic idea that Heisenberg had on Helgoland was this: to ignore completely the electron orbits and take only observable values into account, i.e., the totality of the oscillation frequencies and intensities of the light emitted by the atoms with the spectral lines measured in the spectrograph. In Göttingen, he had already tried to apply this principle to the simplest atom, but at that time this problem had appeared too difficult. Now he was searching for a simpler system by which he could handle the method mathematically. This was the pendulum, which appears in many atoms and molecules as a model of oscillation. It is characterized as anharmonic oscillation.

Heisenberg designated the observable values that play a role in mechanics as "observable" (using the Latin or English word), for instance the position of a particle and its velocity, or the product of velocity and mass, the momentum. Presumably he was recalling his doctoral exam regarding the microscope, where the limit of resolution was evident in the fact that an object smaller than the wavelength of the light illuminating the object could not be "seen" in detail.

If, in this thought experiment, the position and the momentum of an object are to be measured, it clearly depends on whether the position is measured first, and in doing so the object is given a thrust by the light quantum, and then the momentum is determined, or the measurement is carried out in the reverse order. The product of the two measurement processes, or "operations," is different if the sequence of the operations is reversed. The product of position times momentum does not equal the product of momentum times position. The two "observables" are not interchangeable.

Fig. 3.4 Heisenberg, © Werner Heisenberg Estate

In simple numbers, this is not so. Simple numbers are interchangeable or commutable; a X b equals b X a (the commutative law). But if four numbers are arranged quadratically in an array, called matrix A, and rules of multiplication for such a matrix are specified, then the matrices A and B are fundamentally no longer commutable.

Heisenberg carried out this step, although he did not yet know what matrices meant in mathematics. He could calculate the anharmonic oscillator with his new mechanics, but he did not yet know whether his new method was consistent, and in particular whether the law of conservation of energy applied in this new mechanics. So he writes:

[I] work ever more on the question of the validity of the law of conservation of energy, and one evening I had come so far that I could begin to determine…the individual terms of the energy matrix. When the first terms actually confirmed the energy conservation, I felt a certain thrill, so that in the following calculations I kept making miscalculations. For this reason, it was almost 3:00 am before the final result of the calculation lay before me. The law of energy conservation had proven itself valid in all its parts…. In the first moment, I was profoundly shaken. I had the impression of gazing through the surface of the atomic appearances onto a floor of remarkable beauty lying deep below, and I grew almost dizzy at the thought that I was to pursue this wealth of mathematical structures nature had spread out for me down there.

Excited, Heisenberg left his house in the dawn hours and climbed up a crag on the southern tip of Helgoland to await the sunrise.

Paul Dirac, who was a year younger than Heisenberg and was also pondering the physics of the atom around this time, said later that this was "the fundamental idea that occurred to Heisenberg, namely that it was necessary to apply a non-commutative algebra." In retrospect in 1968 Dirac wrote euphorically:

> *I have the greatest reason to admire Heisenberg. He and I were research students at the same time, approximately the same age, and we worked on the same question. Heisenberg was successful, while I failed. At that time, a large amount of spectroscopic data had piled up, and Heisenberg found the proper way to understand them. Thereby he opened up the golden age of theoretical physics, and a few years later, any second-rate student found it not difficult to achieve first-rate results.*

Conclusion in Göttingen

On June 18, 1925, on his return to Göttingen after ten days on Helgoland, Heisenberg stopped over in Hamburg to visit Wolfgang Pauli, who encouraged him to pursue the direction he had taken. In Göttingen, he presented his results to Max Born, and took the idea further. To Pauli's objection that classical mechanics no longer worked, Heisenberg wrote back: "If such a thing as classical mechanics were valid [in the micro-world], one would never understand that such things as atoms existed." On June 24, he wrote Pauli a five-page letter in the second part of which he elucidated his new idea, albeit still in diffident terms:

> *I have almost no wish to write about my own work, because it is all still unclear even to me, and I can guess only approximately how it will turn out. But perhaps the basic ideas are correct after all. The principle is that in the calculation of any values like energy, frequency, and so on, only relationships among verifiable values can be allowed.*

The calculations of the anharmonic oscillator and the formula for the energy values of quantum states follow, as he had analyzed them on Helgoland. "I would be most grateful if you could write to me which arguments work in favor of this formula. Aside from the quantum conditions, I am still not entirely satisfied with the whole schema."

Two weeks after this first letter on quantum mechanics, Heisenberg completed the manuscript of his trailblazing paper "On the Quantum-theoretical Reinterpretation of Kinematic and Mechanical Relations" and sent it in a second letter to Pauli in Hamburg. He was now convinced that he could finally kill the Bohr electron orbits and replace them appropriately. He requested that Pauli send the manuscript back to him in two or three days with "severe criticism" because he would like either to complete or burn it in the final days of his stay in Göttingen. The trust between Heisenberg and Pauli was so close that they sent each other all their papers for critical evaluation before publication. The letter in reply has been lost, but the usually remorselessly critical Pauli must have immediately returned the manuscript with encouraging commentary, for Heisenberg showed it to Max Born in early July and published the paper with Born's approval. In this paper he proposes "entirely abandoning the hope for an observation of the previously unobservable values (like position, or period of revolution of the electron)" and conceding "that the partial agreement of the quantum rules with experience is more or less coincidental." After this settlement with the Bohr quantum conditions, he tries, "to develop a quantum mechanics analogous to classical mechanics, in which only relations among observable values appear."

While Heisenberg began this paper in the first chapter with the definition of the kinematic terms of quantum mechanics, he leaned on Einstein's paper on the electrodynamics of bodies in motion. Einstein, too, had placed value on expressing the equations of electrodynamics in measurable values. Heisenberg closed his paper with the modest observation: "Whether a method of determining quantum theoretical data by relationships among observable values…can be regarded as satisfying…will be realized only through a far-reaching mathematical examination of the methods employed here superficially."

On July 15, 1925, Max Born wrote to Einstein about his brilliant young staff, Heisenberg, Jordan, and Hund: "It requires an effort for me to follow them in their observations…. Heisenberg's new paper, which will soon appear, looks very mysterious, but is certainly correct and profound."

Göttingen had overtaken Copenhagen and Munich in quantum physics. In the following months, Born recognized "suddenly" that Heisenberg's method corresponded to a mathematical matrix calculus that he knew. The "Triumvirate Paper," in which Max Born, together with Heisenberg and Pascual Jordan, worked out a mathematically grounded, complete theoretical mechanics of the atom on the basis of Heisenberg's idea, which the three authors titled "Quantum Mechanics."

The first person to greet Heisenberg's method with jubilation was Pauli, who observed that Heisenberg's mechanics had offered him *joie de vivre* and hope. It was received less jubilantly by the established school of Bohr in Copenhagen, who still hesitated to abandon their model. But news of the new theory spread fast.

Discussion with Einstein

Heisenberg became famous overnight, and the professors at the Berlin stronghold of physics weren't sure what to think of the new quantum theory. Einstein was curious about Heisenberg's "big quantum egg," as he dubbed the theory. He had written to Born's wife: "The new Heisenberg-Born ideas keep everyone breathless. They command the attention and reflection of all those interested in theory.... In the place of gloomy resignation comes a singular excitement to those of us with blood in our veins."

He had reservations about individual points in the new theory, but he did not write to the creator of the idea nor to his friend and frequent recipient of his correspondence Max Born. Instead he wrote to Pascual Jordan, the youngest of the Göttingen people. Heisenberg got to read the letter, of course, and replied on November 16, 1925. He gave the "Honorable Professor" Einstein to understand who the actual originator of the new theory was: "I was very pleased with your very friendly letter to Jordan, and since I feel somewhat responsible for the mischief caused by the new theory, I would very much like to reply to the objections of your letter."

He went into individual points and mentioned that Pauli had succeeded in deriving the Balmer formula (for the spectrum of hydrogen) on the basis of the new mechanics, "and the calculations involved are scarcely longer than in the previous theory. Even if the new theory essayed gives rise to greater complications in calculation, one might nonetheless consider that nature 'does not do it cheaper.'"

He goes on:

I do not know whether the basic premises of the theory are unappealing to you a priori, but it seemed to me that rescue from this piling up of difficulties in recent years of quantum theory is not even possible unless one considers precisely the values observable in general. That in the process, visualizability ("Anschaulichkeit") is completely lost has also seemed very unfortunate to me, and for a time I have not trusted myself at all to publish the thing. But I have eased my conscience with the thought that there would surely be no atoms if our [present] concepts of space-time in very small spaces were even approximately correct.... With heartfelt greetings from Prof. Franck, I remain yours respectfully, W. Heisenberg.

However, the Berlin physicists wished to understand it more precisely, and so in April 1926 Heisenberg was invited to lecture to the assembled luminaries at the Berlin physics colloquium. Heisenberg prepared himself carefully for this encounter with the great world of science and presented his theory, trying to define the unfamiliar new concepts as clearly as possible. After the two-hour lecture and a thorough discussion with the audience, Einstein invited him to his house to discuss these new ideas more comprehensively.

Einstein immediately recognized the essential point: Heisenberg wanted to eliminate electron orbits entirely, even though such tracks were visible in a cloud chamber. Heisenberg described his conversation in his memoir *Physics and Beyond* . He countered that the orbits of the electrons in the atom were not observable; only the transitions between the quantum states of the electrons. The entirety of spectroscopic data of an atom served him as a substitute for the electron orbits. Einstein objected that he could not seriously base a physical theory exclusively on observable values.

Heisenberg responded that Einstein himself had used this principle in the foundation of relativity theory when he defined time as something measured by clocks. According to Heisenberg's recollection, this argument parried Einstein. It might perhaps be of heuristic value to recall what one actually observed. But from a viewpoint on principle it was false to want to base a theory only on observable values. For in reality it is just the reverse. The theory first determines what is observable. If we wished to assert we had observed something, we would have to know the natural laws at least in practice. Only theory, i.e., knowledge of natural laws, permits us to infer the fundamental process from the sensory impression. Heisenberg was very surprised by this position of Einstein's, and the conversation then turned to the interpretation of the ideas of the physicist and philosopher Ernst Mach. Ultimately, the two could not agree, but Heisenberg proposed as a compromise that the mathematical structure of the new mechanics was correct, although how it related to the usual language was not yet established.

In the course of the discussion, a further difference emerged: Heisenberg was speaking about what we know about nature, while Einstein insisted that one had to speak about what nature actually does. This difference is profound: Einstein had the idea that theory had to account for the reality of nature, whereas Heisenberg saw himself compelled by the behavior of atomic systems only to ask that the theory correctly describe our observations. Einstein's conception is that of classical physics of the 19th century: a physical theory describes the reality of natural processes, and permits us to predict the sequence of events exactly. Processes are determined, that is to say, established and unambiguous.

Today we know the Keplerian orbit of a planet around the Sun, so we can calculate where the planet will be located in ten or a hundred years. Since we know Goethe's date of birth, "on the 28th of August, 1749, noon, at the stroke of twelve in Frankfurt on Main," we can calculate the position of the Sun at this time. Likewise, one can determine which constellations were in the sky on the Ides of March, in the year 44, when Caesar was murdered. At the impact of two balls, from the conditions before the impact, we can calculate exactly the path each of the two balls will take after impact.

This determinism is revoked in the quantum world. It is true that in the impact of an X-ray quantum against a stationary electron, the rules of conservation of energy and momentum are still valid, but it is not possible to predict where the impacted electron and the X-ray quantum shifted in its energy will go.

The discussion must have lasted very long. It ended with the question of which criteria for the truth of a theory in physics should prevail – verification by experiment alone, or backed by the simplicity and beauty of the mathematical forms as a criterion of esthetic truth and as a form of economical thinking. The questions raised about the reality of a theory, and about determinism in elementary processes, were, in the years that followed, hotly debated. Often Bohr and Heisenberg stood on one side and Einstein on the other. The paradigm shift in the theory of the atom was a great step forward, which some were not prepared to take.

The Completion of the New Quantum Theory

In October 1925, Wolfgang Pauli wrote enthusiastically to Ralph Kronig in Copenhagen about Heisenberg's mechanics, which, as he says, "gives him new hope," though he was unable to withhold his critique that it would not bring a solution to the puzzle and that it had to be freed from the torrent of erudition coming out of Göttingen. Incensed, Heisenberg, who had come to read the letter, sent a rough "sermon in Bavarian," urging his friend to stop using coarse language. Pauli took the sermon seriously and sat down to read the Triumvirate's paper closely and to calculate as a prime example of the application of the theory the energy levels of the simplest atom, the hydrogen atom.

The Göttingen physicists had not yet been able to solve this problem, but Pauli, with his superior mathematical abilities, arrived at the result in a week. Heisenberg was enthusiastic over the speed with which Pauli had reached this important finding of the new theory. The opponents of quantum mechanics, which at this time included Einstein, were amazed when Pauli succeeded in calculating the energy level of the hydrogen atom using the matrix method.

Even more important for Heisenberg was Paul Dirac's paper published in December 1925 in Cambridge "The Fundamental Equations of Quantum Mechanics." In the summer, Heisenberg had entrusted the manuscript of his first paper on the new mechanics to his host Ralph Fowler, who had passed it on to his doctoral student Dirac, with the question, "What do you think of it?" In short order, Dirac developed an alternative, mathematically consistent formulation of the theory, which established the connection to the classical so-called Hamiltonian mechanics. He sent Heisenberg the galleys of his paper, who read the "uncommonly beautiful paper on quantum mechanics with great interest" and communicated this to Dirac.

The Schrödinger Equation

In March 1926, eight months after Heisenberg's trailblazing publication, another alternative version of quantum physics appeared, the "wave mechanics" of the Viennese Ernst Schrödinger, now a professor at Zürich. On the basis of the idea of the French theoretician Louis de Broglie, Schrödinger conjectured that every particle, including the electron, also displays a wave property. In two communications to the *Annalen der Physik* on the quantization as an Eigen-value problem, Schrödinger extended this conception by showing one could construe the Bohr electron orbits in the atom by requiring that the length of the orbit be a integer multiple of the wavelength, that is, that the electron waves fit within the orbit.

As in a vibrating string, there are oscillations without a nodal point, or with one or more nodes. When Schrödinger presented this idea at the colloquium in Zürich, his colleague Peter Debye remarked that one needed to establish a wave equation for this phenomenon. Schrödinger did this, and this wave equation offered an elegant solution for the hydrogen atom. It permitted calculation of the energy states as Eigen values of the differential equation. The spectral lines emitted by the atom could be construed as beats of the frequencies belonging to the energy states, analogous to the simultaneous sounding of two notes on two strings of a violin where one can hear the beats between the two frequencies.

The results of the calculation agreed with that of the quantum mechanical calculation, but the methods of the Heisenberg quantum mechanics and Schrödinger's wave mechanics, or so-called "undulation mechanics," were entirely different. As Schrödinger wrote, common to both approaches was the understanding, "that in the atomic clusters no outstanding importance at all is to be given to the electron orbits themselves, and even less to the position of the electron and its orbit."

Heisenberg guessed that the contents of Schrödinger's paper must be closely related to quantum mechanics, and Schrödinger saw the connection as well. He was anxious, though, to distinguish his paper from those of the physicists at Göttingen.

In his second communication, Schrödinger quoted Heisenberg's paper and the Triumvirate's paper with a very tortured explanation:

> *At this point, I should not like by remaining silent to ignore the fact that an effort is currently under way on the part of Heisenberg, Born, and Jordan, and several other outstanding researchers [he means Dirac] to eliminate the quantum difficulty which can already point to such noteworthy successes, that it will be difficult to doubt that it contains at least a partial truth. In this trend, the Heisenberg effort is exceptionally close.... His method is so toto genere different that I have not heretofore succeeded in finding the connecting link.... The strength of the Heisenberg program lies in the fact that it promises to yield the line intensities, a question from which we have previously held ourselves quite apart. The strength of the present attempt lies in the leading physical viewpoint which bridges the macroscopic and the microscopic mechanical events.*

With this explanation, Schrödinger recognized the precedence of Heisenberg but stressed that his wave mechanics was more "*anschaulich*" (accessible to imagination) and was on an equal footing with quantum mechanics. He hoped that his theory could be interpreted classically as wave equation, and that he would be able to avoid such unfamiliar new ideas as quantum jump. This new term refers to the fact that electrons in the atom jump from a higher energy state to a lower one stepwise, and the atom emits the difference of the two energies as light quanta.

All those physicists such as Wilhelm Wien or Einstein, who were dissatisfied with the new terminology of quantum mechanics, and particularly with the abandonment of a deterministic world view, saw a way out in Schrödinger's equation. After a lecture by Schrödinger in Munich on July 23, 1926, Wien was enthusiastic and thought that now, evidently, the thesis of quantum jumps had been replaced with something sensible. Heisenberg had traveled from Copenhagen to Munich expressly to hear the lecture and disagreed, as did Sommerfeld, who also heard the lecture. As Heisenberg wrote later, Wilhelm Wien took as a slight that one no longer could steer back under full sail towards the land of classical physics. He wrote to Pauli, "Schrödinger throws everything quantum-theoretical overboard, viz., the photoelectric effect, the Frank collisions, the Stern-Gerlach experiment. Then it is not difficult to create a theory."

Sommerfeld endorsed this statement: "Wave mechanics is an admirable micro-mechanics, but the fundamental quantum puzzles are not solved thereby in the slightest."

Schrödinger in Copenhagen

The experimental physicist Wien's rejection of quantum mechanics following the Schrödinger lecture in Munich depressed Heisenberg so much that he asked for Bohr's help. On behalf of the Danish Academy, Bohr extended an invitation to Schrödinger, who gave his grateful assent "to speak [about the] difficult and burning questions." In the first week of October 1926, he arrived in Copenhagen.

An intense, eight-day conversation developed between Bohr and Schrödinger, in which Bohr attempted to persuade his visitor that his matter-waves were in no way a means of avoiding the quantum jumps in the atom. Schrödinger, on the other hand, was so distraught that he said if one were to stick with this damned quantum jump business, he regretted ever having busied himself with the problem. Bohr's reply characterized the gentleman: "But Schrödinger, we are so grateful to you for having done it."

The adversaries parted on friendly terms in October 1926. Bohr and Heisenberg were convinced they had won the debate, while Schrödinger maintained his aversion to quantum jumps. But already in April 1926, in a letter to Pascual Jordan, Pauli had shown that the two theories were equivalent. Paul Dirac, too, demonstrated in detail the equivalence of quantum and wave mechanics in his paper "On the Theory of Quantum Mechanics" of August 1926.

Born's Probabilistic Interpretation

Max Born proposed a new interpretation of Schrödinger's wave function. Whereas Schrödinger thought his electron waves presented the particles in motion directly, Born offered an alternative conception: the wave function Ψ

was the guidance field that spreads in accordance with the Schrödinger equation, and determines the probability of following a given path. "The motion of the particles follows laws of probability, but the probability itself spreads in agreement with the law of causality." This interpretation was also accepted by Bohr.

The Uncertainty Principle

After Schrödinger's departure from Copenhagen in October 1926, Bohr and Heisenberg were convinced they were on the correct path to understanding quantum mechanics. As Bohr's assistant, Heisenberg lived in a garret room at the Institute. Bohr was busy all day long with administration of the Institute and teaching duties. Heisenberg had to hold lectures and participate in seminars. So, discussions on the physical meaning of the theory often began around 10:00 pm, when Bohr came into Heisenberg's room over a bottle of sherry. Then the discussions went on until midnight.

Bohr's idea was to allow wave and particle to coexist as equally valid, even though this contradicted mathematical logic, whereas Heisenberg's proceeded from his quantum mechanics, and he expected that a consistent interpretation of the physically observable values had to follow, and that in this, there was no freedom. Now began a fierce two-month-long struggle between the two over which was the "correct" interpretation. While for Bohr, the philosophical concepts and their physical meaning stood first and foremost, for Heisenberg, how the theory could describe observation was decisive. Bohr was well-known for often speaking indistinctly to leave a free field for his interlocutors' thoughts. When the topic became particularly difficult, he even sometimes held his hand over his mouth. The endless discussions led to no essential result and exhausted both discussants. Bohr's second assistant, Oskar Klein, remarked that Bohr was "very tired." So Bohr decided to travel to Norway after semester's end, mid-February 1927, for an extended ski vacation.

This decision was liberating for both. Heisenberg, who remained in Copenhagen, was quite happy to be able for once to reflect independently and alone. He wrote to his parents: "In the last two weeks I have undertaken a systematic ordering of ideas for my private use, and now I see clearly what problem I will aim for…. Up to now, I've been too dumb to solve it."

He recalled his conversation in Berlin with Einstein, who had said, "First, the theory determines what can be observed." The trail of an electron in a cloud chamber is not observable directly; only the water droplets along its path are, which are far larger than an electron. The particle's position is ascertainable only approximately, i.e., with a certain imprecision, and likewise its velocity only approximately.

Was there a relation between these two imprecisions? A quick calculation showed that in quantum mechanics the imprecisions are tied together by a relation which we call today the uncertainty principle. The product of the imprecisions cannot be zero; rather it has approximately the value of Planck's constant, h.

As soon as a week after Bohr's departure, in a 12-page letter of February 23, 1927, to Pauli, Heisenberg described his thoughts about the intuitive (*"anschaulich"*) meaning of the mathematically complete quantum mechanics. In it, in eight points he elucidated to his friend what the issue was: "The question of the position of the electron should be replaced with a different one: how can the position x of the electron be determined?" If the position is precisely determined, in that moment, because of the quantum mechanical commutation relation, the momentum p (or velocity) is completely undetermined. Corresponding considerations may be repeated for all canonical pairs of variables for which such a commutation relation holds.

Thereby, it became clear that the position and the momentum of a particle cannot be determined simultaneously with "arbitrary accuracy." Heuristic considerations of the process of measurement with the microscope had already led Heisenberg to this result earlier. Still missing, however, was a quantitative statement of what arbitrary accuracy is supposed to mean in the context of quantum mechanics. Heisenberg could deduce mathematically that the product of the uncertainties of position and momentum had to be greater than Planck's constant, h, divided by 2 times π, that is $h/2\pi$ or $\hbar$. This uncertainty principle means that the inexactness of the observed values does not depend on the insufficient precision of the measuring instruments but is instead a fundamental property of the physical world. The paper concludes with the sentence: "Since all experiments are subject to the laws of quantum mechanics, and thereby to the equation 8.12 [the uncertainty principle], the incorrectness of the law of causality is definitively established by quantum mechanics."

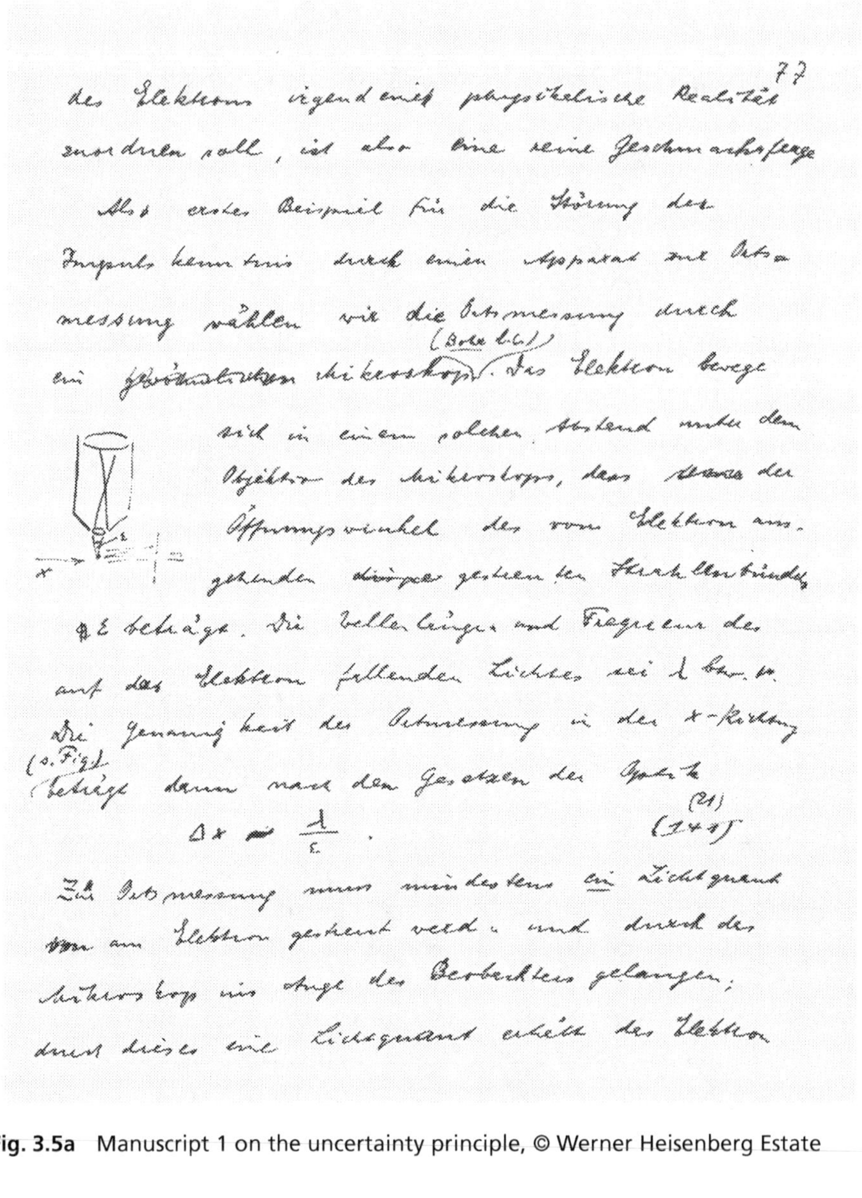

77

des Elektrons irgend eine physikalische Realität zuordnen soll, ist also eine reine Geschmacksfrage.

Als erstes Beispiel für die Störung der Impulskenntnis durch einen Apparat zur Ortsmessung wählen wir die Ortsmessung (Bohr l.c.) durch ein γ-Strahlmikroskop. Das Elektron bewege sich in einem solchen Abstand unter dem Objektiv des Mikroskops, dass der Öffnungswinkel des vom Elektron ausgehenden gestreuten Strahlenbündels ε beträgt. Die Wellenlänge und Frequenz des auf das Elektron fallenden Lichtes sei λ bzw. ν. Die Genauigkeit der Ortsmessung in der x-Richtung (s. Fig.) beträgt dann nach den Gesetzen der Optik

$$\Delta x \sim \frac{\lambda}{\varepsilon} . \qquad (21)$$

Zur Ortsmessung muss mindestens ein Lichtquant am Elektron gestreut werden und durch das Mikroskop ins Auge des Beobachters gelangen; durch dieses eine Lichtquant erhält das Elektron

Fig. 3.5a Manuscript 1 on the uncertainty principle, © Werner Heisenberg Estate

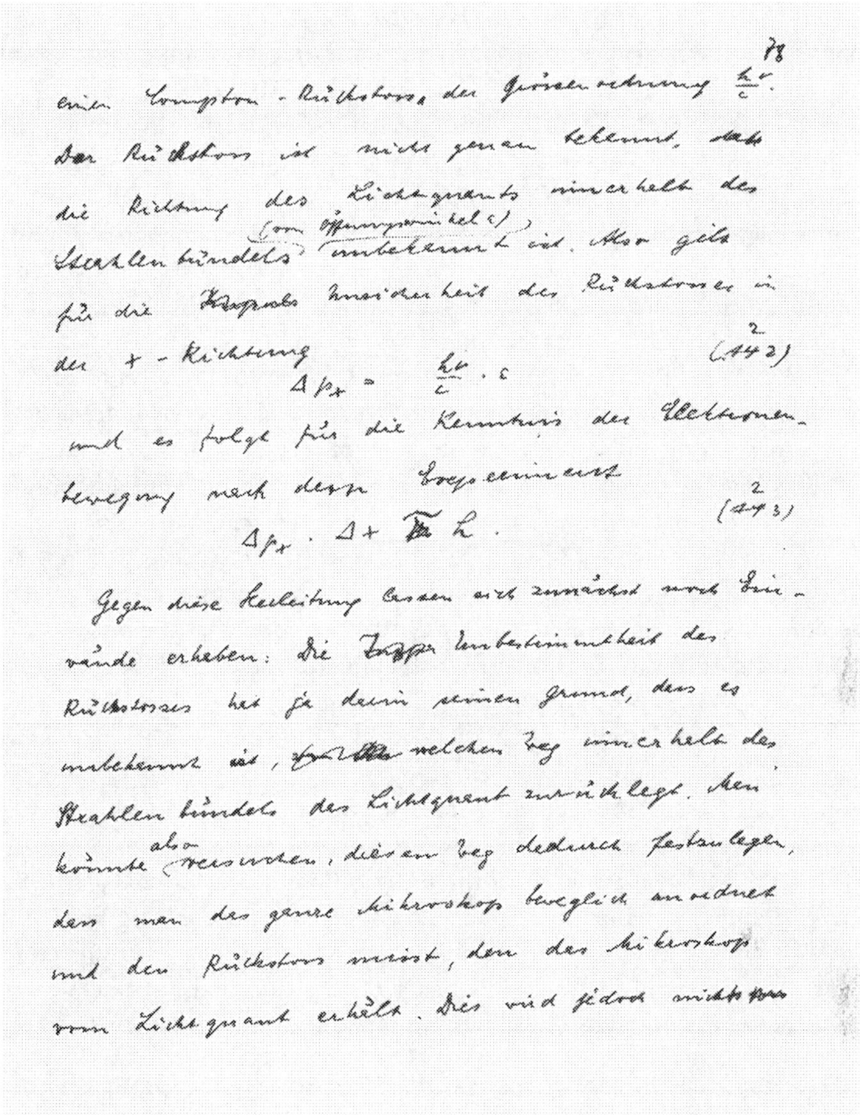

78

einen Compton-Rückstoß der Größenordnung $\frac{h\nu}{c}$.
Der Rückstoß ist nicht genau bekannt, da
die Richtung des Lichtquants innerhalb des
Strahlenbündels (vom Öffnungswinkel ε) unbekannt ist. Also gilt
für die Unsicherheit des Rückstoßes in
der x-Richtung (A4 2)

$$\Delta p_x = \frac{h\nu}{c} \cdot \varepsilon$$

und es folgt für die Kenntnis der Elektronenbewegung nach diesem Experiment (A4 3)

$$\Delta p_x \cdot \Delta x \approx h .$$

Gegen diese Herleitung lassen sich zunächst noch Einwände erheben: Die Unbestimmtheit des
Rückstoßes hat ja darin seinen Grund, dass es
unbekannt ist, welchen Weg innerhalb des
Strahlenbündels das Lichtquant zurücklegt. Man
könnte also versuchen, diesen Weg dadurch festzulegen,
dass man das ganze Mikroskop beweglich anordnet
und den Rückstoß misst, den das Mikroskop
vom Lichtquant erhält. Dies wird jedoch nicht

Fig. 3.5b Manuscript 2 on the uncertainty principle, © Werner Heisenberg Estate

Pauli responded positively to the considerations in Heisenberg's letter, writing, "Daybreak in quantum theory!" In the meantime, Heisenberg had already prepared for publication "On the Visualizable ("*anschaulich*") Contents of the Quantum-theoretical Kinematics and Mechanics," and sent it in turn to Pauli with the customary request for a critical reading and return in a few days.

In mid-March 1927, Bohr returned from his winter vacation. He, too, was impressed by the work Heisenberg had accomplished in the four weeks of his absence. Bohr's second assistant, Oskar Klein, commented, "[I]n those days, [Bohr] lauded Heisenberg like a Messiah." He approved Heisenberg's sending the paper to the editor of the *Zeitschrift für Physik*.

In mid-April, Bohr wrote to Einstein: "This paper surely represents an extremely important contribution to the discussion of the general problems of quantum theory….. The fact that the limits of our ideas coincide so exactly with the limits of our ability to observe, permits us to avoid contradictions."

One month later, Bohr's mood clouded over. He had realized that Heisenberg's paper overlapped the general paper he himself was planning on the conceptual structure of quantum theory, in which he intended to present his principle of "complementarity" with the viewpoint that there were both waves and particles, or "corpuscles." Subsequently, there developed a serious personal crisis between the two. Bohr's qualitative philosophical interpretation collided with Heisenberg's mathematical formulation of the uncertainty principle. It was a conflict based on different conceptions of a*nschaulichkeit* (visualizability), and probably also on the fact that Heisenberg was operating in an area that Bohr considered his own personal territory.

When the galleys arrived from the printer, Bohr asked for changes in the text. Heisenberg declined, but appended an afterword referencing Bohr, and stressing that his (Bohr´s) investigations permitted a deepening and refinement of the present analysis. This paper on the uncertainty principle was Heisenberg's second groundbreaking work, and it spread his fame.

Fig. 3.6 Uncertainty, © Claus Grupen

Einstein's Reaction

While Bohr and Oskar Klein worked out the paper on complementarity at Bohr's country estate, called Tisvilde, Heisenberg was acting in the place of the Institute's director and had time for correspondence. On May 19, he wrote to Einstein that by way of Born and Jordan he had heard of a paper of Einstein's in which he maintained it was after all possible to determine the paths of particles more precisely than the uncertainty principle suggested. He requested galleys of the paper because he was of course anxious to learn Einstein's thoughts and find out if there were new experiments that could determine who was correct, Schrödinger or statistical quantum mechanics. Einstein had in fact submitted a paper to the Prussian Academy of Sciences on May 5, 1927, with the title, "Does Schrödinger's wave-mechanics determine the motion of a system completely, or merely statistically?" Heisenberg thanked Einstein for his reply, and remarked: "If I understand your position correctly, you think that although all experiments yield results as the statistical quantum mechanics requires, later it will in addition be possible to speak of specific trajectories of a particle."

This hope of Einstein's was not to be realized. At least he withdrew the paper shortly thereafter, so that it was not printed. Heisenberg's conviction was nonetheless that, "recent developments in atomic physics have definitively established the invalidity or at least the irrelevance of the law of causality".

Fig. 3.7 Pauli, Heisenberg and Fermi at the Como-Conference, 1927, © CERN, Geneva

Heisenberg's quantum mechanics and Schrödinger's wave mechanics constituted the principal subject of the two great conferences of the year 1927: the conference celebrating the 100th anniversary of the death of the Italian pioneer of electricity, Alessandro Volta, staged with great pomp in September at Como by Mussolini's fascist regime, and the fifth Solvay Conference in October in Brussels, which all the important physicists of the day attended.

4

Impact of the Discoveries

The Fifth Solvay Conference, 1927

The Solvay Conference took place in October 1927 in the Metropol hotel in Brussels. This was the fifth in a series of conferences which the Belgian chemist and industrialist Ernest Solvay had initiated and financed in 1911 at the instigation of Walter Nernst of Berlin. Following his successes as head of a chemical firm, Solvay made a hobby of devising theories of gravitation and electricity and wanted to steer the attention of professionals to his theories. Nernst suggested he convene a conference of important physicists who could discuss current problems of physics. Thus, this series of conferences came into being in which only about 25 scientists of rank and renown were invited to participate. At the conferences in 1921 and 1924, resentment following the war was still so great that no German scientists could take part.

In October 1927, the chosen topic was electrons and photons, and the conference was to deal with the development of quantum theory. In this area, the most important advances had been made in Germany, Denmark and Switzerland, so this time Einstein, Born, Heisenberg, Bohr and Schrödinger were also invited. The chairman of the conference, Hendrik A. Lorentz, had first inquired of Einstein. Of the Göttingen group, because of "originality regardless of person," Einstein had proposed Heisenberg and Franck, or, in case only theoreticians were to be considered, Heisenberg and Born, as well as Schrödinger for his wave mechanics. The new quantum theories were the central topic of the conference, which included all the important physicists of the day.

K. Kleinknecht, *Einstein and Heisenberg*, Springer Biographies,
https://doi.org/10.1007/978-3-030-05264-5_4

Fig. 4.1 Participants of the Solvay Conference, 1927, © CERN, Geneva

In the session on quantum theory under the chairmanship of Hendrik Lorentz, three topics were scheduled on which de Broglie, Born, Heisenberg, and Schrödinger would lecture. De Broglie explained his reasons why particles of matter also had wave properties; Born and Heisenberg represented their theory of quantum mechanics. Born's summary read, "We see that quantum mechanics provides mean values precisely, but cannot predict individual outcomes. The determinism heretofore assumed to be the basis of the exact natural sciences can no longer be accepted as unconditionally applicable."

Heisenberg concluded with the statement: "The true sense of Planck's constant h is thus that it determines the universal measure of uncertainty, which was introduced into the natural laws through wave/particle dualism."

Schrödinger countered that his wave theory was more accessible to the imagination than quantum mechanics.

The following day in the session "Discussion of Suggested Ideas," Niels Bohr spoke about his complementarity principle, by which he established the dualism between wave and particle concepts as the basis for interpretation of quantum theory. In Bohr's understanding, we can view the same natural process in two different ways. Logically, to be sure, the two ways of observing are mutually exclusive, but on the other hand they complete each other, and only together do the two ways of observing yield a complete picture of the process. With this interpretation of the theory, Bohr departed the sure ground of bivalent logic; in describing quantum phenomena, the "either/or" dualism would no longer apply.

Now the debate began in earnest, opened by the combative Einstein. One and a half years before, he had formulated his critique of quantum mechanics in a discussion with Heisenberg in Berlin and had not changed his position since. In his discourse, Einstein had said, "I am aware of the fact that I have not delved deeply enough into the essence of quantum mechanics." He had strong reservations about it; he was especially dissatisfied with the fact that only probabilities of elementary processes could be calculated using this theory. Physics was thus no longer deterministic; it was no longer possible to predict the course of processes from the known initial conditions of a physical system, as had been the case in classical mechanics as well as in Einstein's relativistic mechanics. For him, it was inconceivable that "God played dice." But he was unable to formulate an alternative to quantum mechanics; he had withdrawn his attempt in a paper of May 5, 1927.

In the days that followed, the dispute over the meaning of quantum theory was carried on in lively discussions. Among the participants was a group that wished to preserve traditional determinism and the claim of classical physics to reality. They were known as the "realists." The other group, with Bohr as doyen and spokesman, Born, Heisenberg, and Dirac, represented the view that quantum mechanics permitted only the calculation of probabilities of elementary processes, and that the theory describes only what is known about the events, not how the events "really" proceed. They acquired the name "instrumentalists" because they stressed that we know only what we can observe and measure.

Einstein dreamed up thought experiments meant to refute the probabilistic interpretation and the uncertainty principle. He would present his latest thought experiment to Bohr and Heisenberg over breakfast. On the way to the conference room the problem was defined precisely, and during the day discussions took place among the quantum physicists that led to Bohr's demonstrating to Einstein over dinner in the evening that his experiment could not circumvent the uncertainty principle. Einstein would come to breakfast the following morning with a new thought experiment, which, although more complicated than the previous one, suffered the same fate. This game was repeated several times, until Einstein's friend Paul Ehrenfest accused him of arguing against the new quantum theory just as obstinately as earlier Einstein´s opponents had fought against relativity theory. Ehrenfest described the debate to his students in Leyden:

> *It was glorious for me to attend the conversation between Bohr and Einstein. Like a chess player, Einstein, ever new examples. Virtually a new kind of perpetuum mobile, to breach [Heisenberg's] uncertainty principle. Bohr, out of a dark cloud of philosophical smoke, constantly seeking out the tools to smash it, example after example. Einstein like a jack-in-the-box. Each morning, popping up again, refreshed. Oh, that was delicious. But I am almost unreservedly pro Bohr and contra Einstein.*

For Einstein, his whole world view was at stake. According to his belief, the physical theory described an objective world that existed in the universe independent of us. The theory also permitted prediction of future development from knowledge of the present, momentary state. That this should no longer be possible according to the description of the atomic world was unacceptable for him. He regarded quantum theory as a temporary phenomenon, which would be corrected in the future. To his credo "God does not play dice," Bohr countered, "but it cannot be our task to dictate to God how he should govern the world."

In discussions outside the conference, Heisenberg said to Einstein, "If I understand your viewpoint correctly, you would sacrifice the simplicity of quantum mechanics for the causality principle." And further, "I find it inelegant to ask more than a physical description of the relation among experiments."

Fig. 4.2 House of famous visitors, © Claus Grupen

The two theories, relativity theory and quantum mechanics, became the basis of modern physics in the twentieth century. They were and still are valid in various realms: relativity theory in understanding the universe and quantum mechanics in understanding the subatomic world of elementary particles. To this day, it has proven impossible to merge the two theories.

Impact of the General Theory of Relativity

After the general theory of relativity – with the spectacular confirmation of the deflection of light in the gravitational field of the Sun in November 1919 – had made its triumphal march through the world, nominations for a Nobel Prize for Einstein piled up from all over the world. While in 1919 only five nominations were submitted, the number grew to 8 in 1920, 14 in 1921, and 17 in 1922. In total, Einstein was nominated 62 times. Finally, in 1922, he was awarded the prize for 1921, but remarkably enough not for relativity theory but rather for the explanation of the photoelectric effect from his paper of 1905. Einstein was now intent on deriving additional consequences from the general theory of relativity.

One of these consequences was the red shift of spectral lines in the gravitational field of the Sun, which was mentioned in Chapter 3. In the gravitational field, clocks should run slower, and the wavelengths of light should be increased, i.e., shifted towards the red end of the spectrum. Einstein persuaded the astronomer Erwin Finlay-Freundlich in Berlin to devote himself to this experiment.

Freundlich advocated successfully for the construction of a solar observatory on the Telegrafenberg in Potsdam, and recruited an architect he knew, Erich Mendelsohn, to carry out design and construction of the "Einstein Tower." Half of the construction was financed by the Prussian state, the other half by a German organization called the Albert Einstein Foundation. A telescope with a focal length of 14 meters was installed in the tower. The spectrographs for measurement of the wavelengths were located in a horizontal low-rise building. The wavelengths emitted by a spectral line of the Sun's surface ought to have been greater by two millionths than an identical spectral line on Earth. It soon proved, however, that disturbance effects such as the turbulence of the plasma at the Sun's surface overrode the tiny effect of gravitation. The experiments in Potsdam yielded no results. Einstein did not live to see the observation of the effect, which could not be distinguished from the disturbance effects and measured with any precision until after 1970.

An alternative method for measuring the gravitational red shift was later enabled by a discovery of Rudolf Mößbauer's. In 1958, he found that atomic nuclei, frozen fast at very low temperatures in their crystal lattices, undergo no recoil when they emit a high-energy light quantum in a quantum jump. Such quanta correspond to energetic "hard" X-rays. Thereby, the wavelengths of this radiation from all atomic nuclei of the same kind are exactly equal, and the radiation can be once more taken up or absorbed by an identical atomic nucleus, while the nucleus carries out the reverse quantum jump.

Two physicists, Robert Pound and Glen Repka, employed the Mößbauer effect to measure the influence of Earth's gravity on the energy state of atomic nuclei. They placed a gamma-ray source of the iron-57 isotope at the bottom of a 22-meter high tower. At the top of the tower, they placed another preparation of the same isotope, which was to absorb the gamma radiation from the source at the bottom. Absent the influence of gravity the receiver at the top of the tower would absorb the radiation, since the wavelengths of emission and absorption are exactly equal. But gravity alters the wavelengths of the emission at the bottom; it is greater than the wave-lengths necessary for absorption of radiation in the preparation at the top of the tower. The radiation is not absorbed. To make the wavelengths suitable, Pound and Repka slowly moved the source upwards.

As everyone knows from experience, a fire truck's siren sounds higher when it is moving towards the observer. This is the Doppler effect. The frequency rises as the wavelengths of the sound wave diminish. In the same way, the wavelength of the gamma ray source is shortened in motion towards the absorber at the top of the tower. At a certain velocity, the motion fits the wavelength to the receiver, and we have resonance. The red shift caused by the differential in gravity at a height difference of 22 meters can be measured from the velocity of the source at resonance, several millimeters per second. This result confirmed the value calculated from the general theory of relativity.

Evolution of the Universe and the Big Bang

From the beginning, Einstein recognized the possibility of using his theory of gravitation to calculate the origin and evolution of the universe. The original idea from which he started in 1916 was of an infinitely extended, static universe. The problem was just that such a universe could not be stable: either it had to shrink because of gravity among the stars, or matter would spread in the form of gas throughout endless space.

In order to preserve a static solution of the field equations of general relativity theory, Einstein was compelled to make unrealistic assumptions. But he found a way out. Like the surface of a sphere, the universe was conjectured to be finite but unbounded. Under these assumptions there were solutions to the field equations. To his astonishment, though, he found that these solutions were not static but were temporally variable, contrary to his firm conviction that the universe had to be static, a matter-filled sphere. To preserve this picture, he analyzed the equations and determined that there was a constant of integration he had set at zero but which could have a finite value. In 1917, he incorporated this constant into the equation and named it the cosmological constant. Thereby he could obtain a static universe as a solution and determine its radius as 10 million light years. Later, Einstein called the cosmological constant his "greatest folly."

While Einstein stuck to his conception of a static universe, other cosmologists went in another direction. The Dutchman Willem de Sitter in Leiden and the Russian Alexander Friedmann in Leningrad found solutions to the field equations that described an expanding universe. In September 1922, Einstein reacted to Friedmann's paper dismissively; the result was suspicious and thus false. Eight months later he had to concede that he had been mistaken. In fact, alongside the static solutions, dynamic solutions were also possible. Nonetheless, Einstein insisted on his thesis that only the static solution made physical sense because the currently observable stars moved only slowly or not at all. As long as no experimental data were available that contradicted his thesis, he found the model of an expanding universe "repugnant."

This did not change until the American astronomer Edwin Hubble set about to determine the escape velocity at which galaxies are moving away from us and their distances. The escape velocity can easily be measured from the red shift of spectral lines of starlight. If the stars are moving away, the wavelength of visible light increases and the frequency of oscillation diminishes, just as the pitch of an ambulance siren moving away from us drops. More difficult is the measurement of the distance of a galaxy from us.

For this purpose, Hubble used a certain type of star, the so-called Delta Cephei. In these pulsing stars, we find an exact relation between the absolute luminosity and the duration of the pulsation. If the absolute luminosity is known, the distance of the star and its galaxy can be determined from the apparent brightness in observation from Earth. The smaller the apparent brightness, the further away the star. After six years of astronomical observation, Hubble had measured the distances and escape velocities of 24 spiral galaxies. In his 1929 paper he described his discovery that the further away they are, the faster they are moving away from us, and the more their emitted light is shifted towards the red end of the spectrum of wavelengths. This means that the universe is expanding. Einstein had to realize that his conception of a static universe was incorrect.

The Belgian astronomer Georges Lemaître went a step further. Our particular position is not special; rather, every position in the finite but unbounded universe is equally valid. Every galaxy is moving away from every other galaxy at a velocity proportional to their distance, like two points on the surface of an expanding balloon. If this is so, the galaxies must originally – at the beginning of the world – have been close together. Lemaître even thought that the entire mass of the universe must at that time have been consolidated into the volume of an atom. From the Hubble constant, known very precisely today, which describes the expansion of the universe, it is calculated that this was the case approximately 13.2 billion years ago. At that time, then, the Big Bang occurred.

Dark Matter

The discovery of the cosmic background radiation by Penzias and Wilson in 1965 offered additional evidence of the Big Bang. This microwave radiation strikes Earth from every direction. The temperature of this radiation corresponds to 2.7 degrees above absolute zero, which is at minus 273 degrees Celsius. We believe this radiation to be from the cooled light particles from a distant primeval time, only about 380,000 years after the Big Bang. The radiation reveals to us how the formation of chunks of matter took place. Radiation at this temperature displays tiny fluctuations around one hundred thousandths of a degree. From this we can conclude how much matter of the kind we are familiar with was present at the time – it amounts to four percent of the total mass of the universe. The rest, whose nature we still do not know, appears to consist of non-luminous, "dark" matter, plus, as has emerged in recent years, mysterious dark energy.

The existence of dark matter emerges from the observation that the stars at the outer edges of spiral galaxies circle the galaxies' centers at a higher velocity than would be calculated on the basis of Kepler's laws of celestial mechanics. Therefore there must be more matter at the center of the galaxy than the totality of visible stars can account for. So far, it is unknown of what massive but non-luminous components this dark matter consists. But whatever it is, it constitutes one quarter of the mass of the universe.

The nature of dark energy is even more mysterious, its very existence uncertain. The hypothesis is based on the observation of supernovae of a certain type, Ia, whose absolute luminosity is the same in all cases, so that their distance can be determined from their apparent brightness. When the escape velocity of these supernovae is measured from the red shift, it is seen not only that the further these galaxies are away from us, the faster they are receding, as Hubble had found, but also that the escape speed of these galaxies accelerates. Such a phenomenon can be described with the field equations of general relativity theory by Einstein's cosmological constant, if this is positive. The value of the constant, with which observations agree, is tiny: 10^{-17} grams per cubic meter. Why the constant has this value – Einstein had originally set it at zero – is entirely unknown.

Black Holes and Supernovae

Furthermore, it follows from the general theory of relativity that a very great mass can permanently capture light, i.e., that there can be "black holes." As a soldier in WWI, the astrophysicist Karl Schwarzschild found the first theoretical evidence for this. He had read Einstein's paper and tried to use the theory on a fluid, non-compressible globe. The concept that the material of such a globe could not be compressed is contradicted by the reality of stars, which increase in density towards their centers. Schwarzschild found a solution to this problem in the context of gravitation theory. But the theory broke down for the innermost layers of the Sun: there was no solution for the sector within 3 kilometers from the center. This radius is named the Schwarzschild radius. Gravitation in this realm is so strong that nothing can escape it, not even light. The interior is absolute black.

No one was interested in this paradox until Fritz Zwicky and Walter Baade hypothesized in 1933 that enormously dense stars consisting only of neutrons could form out of the remains of supernova explosions. The uncharged neutrons could be packed together as tightly as the building blocks of atomic nuclei. Supernovae result from stars at the end of their life cycle, when their nuclear fuel is "used up" in nuclear fusion. In most stars, hydrogen is first fused into helium, then helium into carbon, until finally iron is created. If the mass of the star exceeds that of eight solar masses, the star collapses at the end of its life into a neutron star or a black hole.

The existence of such exotic phenomena was first taken seriously in 1963 when the Dutch astronomer Maarten Schmidt from the California Institute of Technology observed the visible light from a source of radio radiation known as 3C273 with the mirror telescope on Mount Palomar. From the red shift of the optical spectral lines, he determined its distance, from the radio signal, its luminosity. He confirmed that the object lay far outside the Milky Way, but possessed an enormous luminosity nonetheless. It could not have been a "normal" star. He named the object quasi-stellar object, or "quasar." The energy a quasar emits is billions of times that of a normal star. A few years later, Boris Zel'dovich suggested that the mechanism of this energy might be the absorption of gas and stars by a black hole. The matter collects in disk form around the center, spinning ever faster, and losing energy through radiation until it disappears behind the Schwarzschild radius.

Such a very massive black hole exists at the center of our Milky Way, as Reinhard Genzel of the Max-Planck Institute for Extraterrestrial Physics in Garching discovered. Genzel wanted to "look" into the center of the Milky Way, which is obscured by interstellar dust clouds. He developed a technology by which he was able to observe the center by its infrared radiation. Near this center he observed a series of fixed stars and determined that these stars were not at all "fixed" but rather in motion. After 15 years of observations he could demonstrate that these stars moved in elliptical orbits around a black hole, like planets in their Keplerian orbits around the Sun. From the orbital parameters, it emerged that a massive object of 3.8 million solar masses holding the stars in their elliptical orbits must lie at the center.

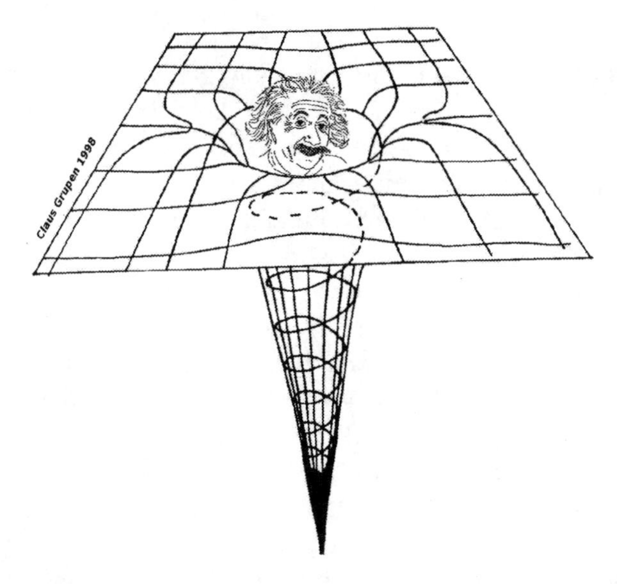

Fig. 4.3 Einstein in a Black Hole, © Claus Grupen

Gravitational Waves

In 1916, Einstein calculated the emission of gravitational waves through accelerated masses as one of the consequences of the general theory of relativity. Einstein thought it impossible that this tiny effect would ever be observable. In 1936, he even believed he had found proof that gravitational waves cannot exist, as he wrote in a letter to Max Born. Whereas in the Newtonian theory of gravity, the force effect of a mass on other, distant masses occurs simultaneously, in relativity theory this is not possible because the effect of the gravitational field in a distant position arrives only after a delay determined by the speed of light.

In electrodynamics there are positive and negative charges, from which a dipole can be formed. As an antenna, an oscillating dipole can emit electromagnetic waves. By contrast, in the case of gravitation, there are only positive masses, and no dipole. The antenna for emission of the waves then is similar mathematically to a quadrupole. Every kind of accelerated mass emits these waves, although their effect is so tiny that demonstration is enormously difficult. These waves, like the electromagnetic, are transversal, and space at right angles to the direction of propagation is itself changed. If a gravitational wave rushes through the detector, the distance between two bodies perpendicular to the direction of propagation of the wave increases or decreases briefly.

The anticipated length change was so small that only a few physicists ventured to devise an experiment. The first was Joseph Weber of the University of Maryland. He constructed very heavy aluminum cylinders, on which he measured the tiniest vibrations. In order to minimize disturbances caused by the thermal motion of the atoms, the cylinders were cooled to the temperature of liquid nitrogen. In 1969, he found simultaneous vibrations in six detectors of this type, but the physics community took the observation to have been an accidental coincidence. It was also unclear what huge cosmic event could have brought about a gravitational wave of such strength.

An alternative, more sensitive method of measurement of gravitational waves was suggested the following year by Rainer Weiss of the Massachusetts Institute of Technology. The demonstration was to be made with a Michelson interferometer. Around 1900, with such an apparatus, Michelson and Morley had shown that the speed of light in the direction of Earth's movement and perpendicular to it were identical; in other words, that the idea that light – the electromagnetic wave – propagated in a material ether could not be correct. Instead of a conventional light source, the gravitational wave detector would work with a laser-beam. The beam is split through a semi-transparent mirror M into two beams perpendicular to each other. Both beams are reflected back by mirrors at the end of their spectrometer arm and meet again at M where the two electromagnetic waves overlap – they "interfere." If the length of the two beam paths is the exactly the same, the waves are added to each other and yield a bright picture. If one of the beam paths is shorter by half a wavelength, the waves cancel each other, and the picture is dark.

A gravitational wave rushing through the detector alters the distances in the two arms of the interferometer, perpendicular to the direction of motion of the gravitational wave, and the interference pattern ought to show that.

At first, the American group did not receive the resources necessary to build a large interferometer. A Munich group led by Heinz Billing had more luck. In 1975, they were able to construct an interferometer with an arm length of 3 meters, and in 1983 one of 30 meters. With these, they succeeded in overcoming the technical difficulties, viz., stability of the lasers, a vibration-free mirror mount, and mechanical stability of the interferometer arms.

In 1980, the American group at the California Institute of Technology received the means to construct a 40-meter interferometer, but it soon became clear that the required sensitivity could be achieved only with kilometer-long arms. The German group together with physicists from Glasgow and the American group proposed such projects. But the resources in Germany sufficed only for a 600-m project, Geo600, which was approved in 1994, and with the participation of British physicists was built and completed by 2005 near Hannover under the direction of Karsten Danzmann. In this experiment, the sensitivity of the technology was increased so much that a change in length of one thousandths of the diameter of an atomic nucleus was detectable.

Though the American project LIGO received initial funding in 1988, it struggled with numerous difficulties. The project did not get going until 1994 under new management. In 1997, the two 4-kilometer interferometers in Livingston, Louisana, and Hanford, Washington, began operation, without spectacular results. Even the reconstruction from 2007 to 2009 as "Enhanced LIGO" was not able to gather signals of a gravitational wave. Not until the group around Danzmann in Hannover installed their critical improvements in the American LIGO detector in the four years from 2011 to 2015 was the sensitivity increased to the point that success became achievable. The more stable high-performance laser, the improved vibration-free mirror mount, and a laser technology called "squeezed light" led to a breakthrough for the detector, now named "Advanced LIGO."

On September 14, 2015, the first day of regular operation in the United States, the improved detector was monitored from Hannover. At 10:50 Central European Time, the supervising physicist registered a signal of surprising clarity. The length change of the interferometer arms at the two positions, Livingston and Hanford, ran like a wave lasting approximately 0.2 seconds. The frequency of the wave was 30 Hertz at the start and rose steadily before finally reaching 300 Hertz. Then it stopped suddenly. The signals arrived in Hanford seven milliseconds later. The 4-kilometer long interferometer arms of both LIGO detectors changed their lengths by a tiny amount, one thousandths the diameter of a proton. The size of the building block of the atomic nucleus is one billionth of a micrometer.

The LIGO researchers had quickly struck gold. They had analyzed possible events beforehand, and this signal corresponded exactly to their simulation of a spectacular event: two black holes orbiting each other, emitting gravitational waves and thereby losing energy. They orbited ever faster around each other before finally merging, forming a single black hole. An exact comparison of the data with the simulation yielded further that the two black holes had masses of 36 and 29 solar masses, respectively, and that the black hole created had a weight of 62 solar masses. The sum of the masses of the two merging black holes was 65 solar masses. So the remainder of three solar masses is, according to Einstein's equation $E=mc^2$, converted into the energy of the radiated gravitational waves. This energy is 100 times greater than the radiation energy of all the stars in the universe put together.

The gravitational wave event was dubbed with the name of its date, GW150914. A second, similar event, GW151226, was registered on December 26, 2015. Thus, 100 years after the formulation of the general theory of relativity, a final cornerstone was found, undergirding Einstein's theoretical edifice.

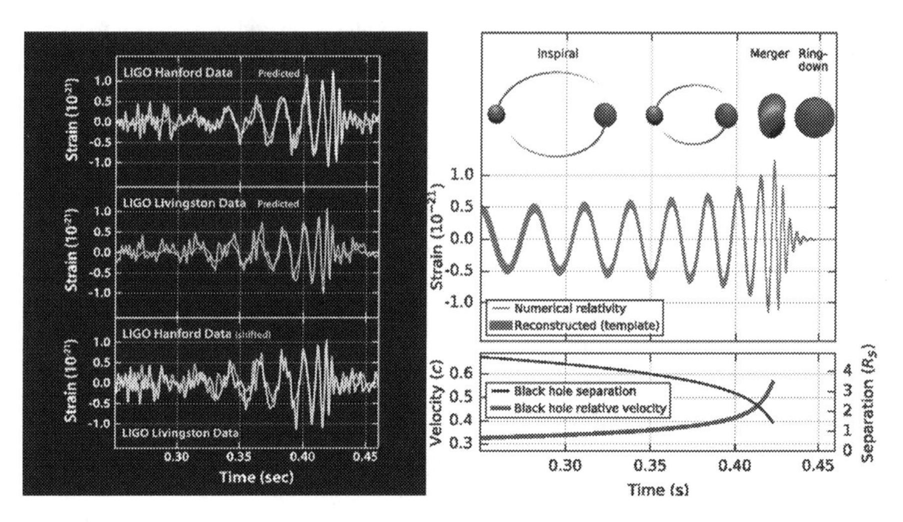

Fig. 4.4 Signal of gravitational wave GW150914, LIGO collaboration (left); formation of the gravitational wave GW150914 (right); "strain" signifies the relative change in length of the spectrometer arms during the passage of the gravitational wave

Knowledge of the origin of the universe has completely changed our world view. Niels Bohr wrote: "Mankind's horizon has been widened immensely through Albert Einstein's work, and concomitantly, our picture of the universe has attained a coherence and harmony one could heretofore only dream of."

Heisenberg in Leipzig

In the summer of 1927, while Heisenberg was Bohr's assistant in Copenhagen, he was offered a professorship in Leipzig for the second time. Debye had moved from Zürich to Leipzig and wanted Heisenberg to join him there. Conversely, the ETH in Zürich needed to search for a replacement for Debye. This faculty also decided in favor of Heisenberg, and at the Como Conference, the invitation of the ETH Zürich was extended to him. On his return trip from Como, he visited the president of the ETH and the physics institute where Paul Scherrer and Hermann Weyl briefed him on the working conditions, and expressed their interest in working with him. But Debye in Leipzig did not give up, and pressed the ministry in Dresden to bring the negotiations to a successful conclusion, for: "Heisenberg's reception in Como has once more shown most clearly that in the general view, the focus of the modern developments will be moved to Leipzig."

In November, Heisenberg decided for Leipzig, and the ETH Zürich in his place appointed Wolfgang Pauli, who likewise accepted immediately. Heisenberg left Copenhagen and took up his new position as full professor at the University of Leipzig. He delivered his introductory lecture on February 1, 1928.

Fig. 4.5 Heisenberg at his introductory lecture in Leipzig, 1928, © Werner Heisenberg Estate

From 1928 on, Heisenberg, as a young professor at the University of Leipzig, was the great attraction for the most gifted students and young scientists from Europe and America. The most interesting field of work at that time was the newly discovered quantum mechanics, and where better to learn it than with its young discoverer himself? Heisenberg set the tone, the "spirit of Linnéstraße." His most famous doctoral students were Felix Bloch from Switzerland (doctorate in January, 1929), Eduard Teller, from Hungary,

Rudolph Peierls, and Carl-Friedrich von Weizsäcker. Bloch came to Leipzig in 1927. Later, he reported,

> *I was his first student, and so I got a lot of time with him. I participated in his seminars right away, and assumed a very close relationship with Heisenberg – I admired him tremendously.... Heisenberg said to me in a friendly way: one ought to have a look at what will come of the electron theory of metals in quantum mechanics.*

Thus, under Heisenberg's guidance, Bloch turned the attention of quantum mechanics to solid matter, thereby establishing theoretical solid-state physics. In his assessment of the dissertation, Heisenberg wrote: "In my opinion, Bloch's paper represents a very valuable contribution to the theory of metals. As a sound basis for further examination it will prove far more applicable than previous theories."

Bloch, who took his habilitation with Heisenberg in 1931, later wrote about his time with Heisenberg in Leipzig:

> *[They] belong to the happier times before these events [i.e., his expulsion in 1933 by the Nazis]. Many of them represent utterly mundane and anything other than professional conversations and excursions, in ski huts in the Bavarian Alps, or under other recreational circumstances. They are no less precious to me than our discussions about physics.*

The friendship continued after the war, too. When Bloch received the Nobel Prize in physics in 1952, he wrote to Heisenberg: "During all the eventful years since I saw you last, I have never lost the sense of profound attachment for all that you have given me."

Heisenberg was the magnet that attracted everyone, not only the students and doctoral candidates but also young scientists. The most gifted of the younger generation of physicists crowded his seminar about the structure of matter. The guest list from abroad in those years was comprised of more than 50 well-known names.

Fig. 4.6 Heisenberg with his students in 1930. *Left to right:* Giovanni Gentile, Rudolf Peierls (front), George Placzek, Gian Carlo Wick, Werner Heisenberg (front), Felix Bloch, Viktor Weisskopf, and Fritz Sauter, © Werner Heisenberg Estate

We get an impression of the "Spirit of Linnéstraße" from the letters of the young Italian Ettore Majorana, whom Heisenberg had admitted in January 1933 to his institute. He worked there on the nuclear forces and later on the mass-less neutral particles, whose existence Pauli had postulated, to "rescue" the conservation of energy in beta decay, the neutrinos. In radioactive beta decay, an electron, the beta particle, is emitted from the atomic nucleus, and it has the appearance of violating the principle of conservation of energy. By conjecturing an invisible particle, the neutrino, emitted simultaneously with the electron, Pauli demonstrated the possibility of preserving the conservation of energy.

In his book *The Disappearance of Ettore Majorana*, the writer Leonardo Sciascia describes the atmosphere at the institute:

> *His meeting with Heisenberg, we believe, was the most important event in the life of Ettore Majorana (1906-1938), more on personal than scientific grounds. Not surprising: on the basis of what we know from documentation of his life, despite all we do not know, we get the impression of a more extensive, important encounter.*

On January 22, 1933, Majorana wrote his mother: "I was very warmly received at the physics Institute. I had a long discussion with Heisenberg, who is an uncommonly polite and sympathetic person."

And one month later:

At the last 'colloquium,' a weekly gathering of around a hundred physicists, mathematicians, chemists, etc., Heisenberg spoke on nuclear theory, and praised me highly for a paper I have written here. Because of our many scientific discussions and a few games of chess, we have become pretty good friends. The occasions for these were offered by the reception he hosts every Tuesday for professors and students of the Institute for Theoretical Physics."

Sciascia sums up his impression of Heisenberg's effect thus: "As we believe we see in retrospect, the reason lay in the fact that Heisenberg lived the problem of physics; his research stood in a broad and dramatic context of thought. He was.– to put it tritely – a philosopher."

In January 1933, the political situation grew dark. The National Socialists seized power. Many of Heisenberg's colleagues and students prepared to leave Germany. Nevertheless, he and his friends among the physicists enjoyed an untroubled ski vacation during the Easter vacation at a mountain lodge at Grosser Traithen, a mountain in the Bavarian Mangfallgebirge. In addition to Heisenberg and Bohr, the group included Bohr's son Christian, Felix Bloch, and Carl-Friedrich von Weizsäcker. Rinsing the dishes one evening, Bohr came to the astonishing realization that even with dirty rinse water and dirty towels for drying them, it was possible to get the dishes and the glasses clean in the end. And it was just the same in physics: we have vague terminology and limited logic, but even so we can ultimately make lucid statements about nature.

Fig. 4.7 Skiing with Felix Bloch, 1933, © Werner Heisenberg Estate

"German Physics"

Following the National Socialist seizure of power and Hitler's assumption of the Reich chancellorship in 1934, the Nazi ideologues among the physicists, in particular the Nobel Laureates Philipp Lenard at the University of Heidelberg and Johannes Stark, president of the Physical-Technical Reich Institute and the German Research Foundation, began to polemicize against the modern extension of physics, relativity theory, and quantum mechanics. Stark wished to assume responsibility for physics and reorganize it. The "German" or "Aryan physics" had exactly the same contents with respect to classical physics. Relativity theory was rejected as "Jewish," and likewise quantum mechanics. Lenard and Stark were isolated among their scientific colleagues; Max von Laue prevented Stark's admission to the Prussian Academy.

But they had backing in the National Socialist Party and its satellite organizations. Heisenberg, on the other hand, always stressed in his lectures that Einstein's relativity theory was the obvious basis of further research. Heisenberg was especially clear in his lecture to the convention of natural scientists on September 17, 1934, in Hannover called "The Changes in the Foundations of the Exact Natural Sciences in Recent Times." He highlighted the fundamental importance of relativity theory and quantum mechanics and corrected the "distortions arising from the quarrels of the opinions of the day." He stressed, too, the role of theoretical physics, which "in recent days has sometimes been presented askew to the general public in Germany."

This defense of the new physics ostracized by the representatives of "German physics" was deemed abroad as a courageous step. Wolfgang Pauli wrote to Heisenberg from Zürich: "Your lecture published in Natural Sciences (Journal "Naturwissenschaften") has evoked in me great enthusiasm as usual, for its contents as well as its tactics. Congratulations are in order!"

In 1937, the organ of the SS, *The Black Corps*, published an article titled "White Jews in Science." In it, Heisenberg was attacked directly as a defender of Jewish scientists, and as proconsul of Judaism in German intellectual life. Stark characterized Heisenberg as "the Ossietzky of physics," a dubious comparison if one knew that the Nobel Peace Prize Laureate Ossietzky was badly treated and tortured in a concentration camp and died of the aftereffects. A party functionary wrote to the chief ideologue of the National Socialist Party, Alfred Rosenberg, that the concentration camp was doubtless the proper place for Herr Heisenberg. Heisenberg's reputation was great enough, however, to ward off such attacks.

In January 1937, at one of the many evening musical gatherings in Leipzig, Heisenberg finally met the love of his life. She was the young Elisabeth Schumacher, who was completing an apprenticeship with a bookshop. It was love at first sight, and the two married the following April.

Fig. 4.8 Elisabeth and Werner Heisenberg, 1937, © Werner Heisenberg Estate

Heisenberg had a talent for attracting and encouraging students. But the flowering of Leipzig physics had been threatened since the Nazi seizure of power in 1933 and then brought to an end as the Jewish members of the group turned their backs on Germany. Heisenberg grew lonely.

Einstein as Teacher

Einstein, by contrast, was in every respect a lone fighter, a "Steppenwolf" (lone wolf), as he characterized himself, in reference to Hermann Hesse's novel. This was demonstrated again and again in Einstein's lack of interest in teaching and in building up a community of students. On the contrary in Zürich, Prague, and Berlin he ensured himself maximum freedom for research. His position at the Academy in Berlin was tailored expressly to his needs. He had no teaching duties, no office at the Academy, no secretary. Regular hours were restricted to the physics colloquium at the university, and the sessions of the mathematical-physical class of the Academy.

Now and later in the United States, Einstein never supervised a doctoral student. For the most part, he worked alone at home, and there led a solitary life, always intent on not wasting time on trivial things. For him, transmitting his ideas in physics to younger students numbered among the trivial things. Thus, there exist only a handful of papers, written together with others. An example is the paper with Podolsky and Rosen on the so-called EPR paradox, which deals with the quantum mechanical entanglement of two photons – the "spooky action at a distance," as Einstein called it – between these in quantum mechanics. This paper was intended to demonstrate a paradox in quantum mechanics that Einstein held to be impossible.

Impact of Quantum Mechanics

One immediate consequence of the fifth Solvay Conference in 1927 was the increasing recognition of the importance of quantum mechanics and the uncertainty principle, and the first Nobel Prize nominations in this area were submitted to Stockholm. But only when the suggestions of Planck, Bohr, Einstein, and Pauli came in did the Nobel committee arrive at a decision. The prize for 1932 went to Werner Heisenberg alone, "for the creation of quantum mechanics, whose application has led, among other things, to the discovery of the allotropic form of hydrogen." The prize for 1933 was split between Erwin Schrödinger and Paul Adrien Maurice Dirac "for the discovery of new, productive forms of atomic theory," i.e., alternative forms of quantum mechanics. The ranking of the relative importance of the contributions of the three scientists can be inferred, for example, from Wolfgang Pauli's proposal (which we publish here). He deemed first that Heisenberg's paper preceded Schrödinger's, and second that Heisenberg's creation was the more original. Heisenberg's second great achievement was the establishment of the uncertainty principle, and the recognition of its importance. The proposal letter, translated from German reads:

Zürich, January 29, 1932

Prof. Dr. W. Pauli

To the Nobel Committee for Physics

Stockholm, 50, Sweden

Honored Sirs,

In response to your kind invitation to submit a proposal for the Nobel Prize in Physics for the year 1932, I am pleased to inform you that in my opinion Dr. W. Heisenberg, Full Professor of Theoretical Physics at the University of Leipzig, is first to be considered for this prize. In this regard, most important are his two fundamental papers, "On the Quantum-mechanical Reinterpretation of Kinematic and Mechanical Relations," Zeitschrift für Physik, 33, 879 (1925), and "On the Intuitive Contents of Quantum-theoretical Kinematics and Mechanics," Zeitschrift für Physik, 43, 172 (1927). The first paper lays the foundation for the so-called "matrix-mechanics," which establishes the historically first mathematically precise form of modern quantum-mechanics, while in the second the "Uncertainty Principle" is expressed for the first time, in accordance with which an exact determination of both a particle's position and its momentum is not possible, since the product of the uncertainties of position and momentum has an immutably small lower limit set by Planck's constant, h. It is doubtless unnecessary to cite the many theoretical consequences, or the proposals for experimental research in detail that flow from matrix-mechanics, the substantial importance of the Uncertainty Principle for the consistent interpretation of experience with the help of modern quantum-theory, as well as the many other papers by Heisenberg, in which he has substantially advanced quantum-theory and our understanding of atomic structure.

Conversely, it may be appropriate to mention the number of other researchers in the development of modern quantum-theory and their relation to Heisenberg's achievements. A contribution similar in form to Heisenberg's is Schrödiger's eponymous formulation of the wave-equation, and the wave-mechanics developed by him therefrom, which was thereupon combined with Heisenberg's matrix-mechanics into the uniform gestalt of modern quantum-theory. If I propose Heisenberg before Schrödiger for the Nobel Prize, this is in consideration of the two following circumstances. First, Heisenberg's matrix-mechanics preceded Schrödiger's work. Second, Heisenberg's creation must be regarded as the more original, since Schrödiger's ideas emerged from those of de Broglie (already a Nobel Laureate). With respect to Heisenberg's second great achievement, the statement of the Uncertainty Principle, it is true that Bohr later improved, simplified, and enhanced it. But Heisenberg's contribution of having for the first time recognized the principle, as well as its substance and importance is not thereby reduced.

So I believe that, at the present time and from every point of view, Heisenberg most fits the requirements which, in accordance with the statutes of the Nobel Foundation and the intentions of its founder, are to be expected of a Nobel Laureate.

In highest esteem,

Very sincerely yours,

W. Pauli

Philosophical Consequences

The explanation of quantum mechanics, as realized in the Copenhagen interpretation, also impacted philosophical questions of epistemology. In a lecture entitled "Epistemological Problems in Modern Physics" to the Kant Society in Kiel in the summer of 1928, Heisenberg stressed that in earlier times the natural sciences had been closely tied to philosophy, and that every important natural scientist – Democritus, Aristotle, Kepler, Newton – had also been a philosopher. He regretted to say, however, that today the two disciplines had become alienated from one another, due in part to the fact that in general formulations of problems it is often impossible to distinguish clearly between observer and object. Then, to this philosophical audience, he explained his uncertainty principle, that position and momentum of microscopic objects cannot be determined with arbitrary accuracy. He went on:

> *This observation implies an interaction between observer and object, which alters the object.... Precise knowledge of the velocity precludes precise knowledge of the position: it is complementary to it. Or: causal description of a system is complementary to a space/time description, since for a space/time description observation is necessary. [What Heisenberg is expressing here in abbreviated form is that if we wish to describe the position and point in time of an event, we must make an experimental observation of it.] If we disturb the system, we can no longer track its causal relation precisely.*

As a general philosophical comment, he added:

> *It may be that Bohr's concept of complementarity also sheds light on the mind/body duality. The natural scientist today will assume that knowledge of a mental event is complementary to knowledge of the corresponding physical event since the two kinds of knowledge are mutually exclusive.... For in order, say, to determine the chemical processes of brain cells, it is necessary to disturb the organism such that there can no longer be any question of mental events.*

Quantum mechanics requires a new conception of reality. In the nineteeth century, the reigning idea was that nature existed independent of our observation of it, that its reality lay beyond human awareness. Observation of the atomic world compels us to recognize that there is an interaction between observer and object, that the object is altered by observation, and that we can gain knowledge of physical values of the objects only if we measure them. The picture of an atom now was no longer Bohr's planetary model but rather an atomic nucleus surrounded by electrons, the probability of whose positions resembled a cloud.

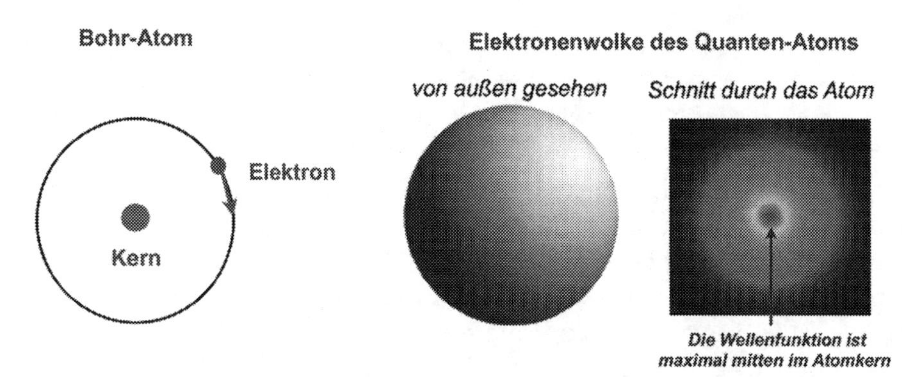

Fig. 4.9 The new model of the atom, left: Bohr model with nucleus and electron; right: quantum atom with electron cloud and cut through atom

Applications of Quantum Mechanics

Immediately following the final formulation of quantum mechanics, the application of the theory to all problems of atomic, molecular and solid-state physics began. Heisenberg himself provided the impetus for treating solid bodies in working out his theory of ferromagnetism. His first doctoral student, Felix Bloch, established the theory of solid-state bodies by applying quantum mechanics to periodic structures of atoms. Soon, quantum mechanics made its triumphal march through the world of physics, and countless applications followed. Today, more or less two-thirds of all industrial production rests on quantum mechanics. Anyone who watches television, listens to the radio, or gathers information from the Internet on a computer, anyone who uses a CD player, makes calls with a cell phone, shoots with a digital camera, scans a document, or uses a laser-pointer, is using quantum mechanical technologies. The same is true of X-ray machines, computer tomography (CT) and magnetic resonance imaging (MRI) scanners, and cancer-therapy with X-rays. Encryption of transmitted data can also be achieved with quantum-cryptography.

Magnetism

Heisenberg's first interest at Leipzig was a theory of ferromagnetism. In question was the phenomenon that the metals iron, cobalt, and nickel within a magnetic field applied from outside amplify the field by several orders of magnitude. This occurs because the electrons in the atoms of these materials align themselves to the external field as elementary magnets. Classical theory could not explain this. But when we take the quantum mechanical exchange forces into account, the electrical forces between the electrons in a crystal lattice are strong enough to affect a collective alignment of the elementary magnets and a huge amplification of the external magnetic field.

To achieve this, as Heisenberg demonstrated, each atom in the crystal lattice must be surrounded by at least eight neighbors, and the electrons responsible for the magnetism must be located in the third shell of the electron cloud. Iron, cobalt, and nickel fulfill these two conditions.

Semiconductors, Integrated Circuits, and Computers

All electronic devices depend on semiconductor technology today. While electrical conductors are constructed from metals such as copper or aluminum, and many materials do not conduct current at all and are useful for manufacture of insulators, there is a group of materials such as silicon or gallium arsenide whose electrical conductivity can be controlled externally with electrical potential. Therefore, these substances can serve as switches and memory storage for information. The characteristics of semi-conductors can be determined with quantum mechanical calculation. The first such switch was the transistor, and technical development makes it possible today to house billions of such transistors or memory elements on a silicon chip the size of a postage stamp. These integrated circuits (IC) and memories constitute the foundation of every computer, laptop, tablet, and cell phone.

Quantum Computer, Quantum Cryptography

Quantum mechanics also makes possible a new method of building computers with the help of quantum conditions. The development of such computing machines greatly increases computing speed because quantum logic permits the simultaneous execution of computational operations. A more advanced development is the use of quantum techniques for encoding data transfer. If data are encoded in this way by the sender, only the recipient can decode them. He or she is also able to determine if a third person is attempting to eavesdrop on the transmission.

The Laser

The light sources we were familiar with before 1900, the Sun, candles, oil lamps, light bulbs, etc., were hot bodies emitting heat radiation and light. The color we see corresponds to the temperature of the radiating body. In the case of the Sun, the yellow coloration corresponds to a surface temperature of 6,000 degrees Celsius, and a light wavelength of about 500 nanometers. But sunlight also contains other wavelengths (i.e., colors), from ultraviolet short wavelength light to infrared heat radiation.

In the transition between two quantum states of an atom, on the other hand, light of only one wavelength is emitted. An atom in a state of excitation, whose energy is higher than that of its ground state, can pass over into the energetically subjacent state: either spontaneously (randomly) during emission of a light quantum, whose energy corresponds to the difference between the two states, or "stimulated" by incident light. If a large number of atoms is successfully transferred into the same excited state by supplying the atoms with enough energy in a mirror array (the resonator) – the process known as optical pumping – it is possible to stimulate all the excited atoms to simultaneous quantum jumps by irradiation. Then we have Light Amplification by Stimulated Emission of Radiation, or a LASER. A very intense light of one color is emitted, and the photons all fly in the same direction along the resonator axis. This very intense, monochromatic, focused light existed first as a red light source, then also as green and blue. There are countless applications, including reading a CD, data storage on a CD, surgical incision of the cornea, cutting massive work pieces, transmission of data and voice along fiber optic cables, laser printers, laser pointers, geodesics, application in the laser interferometer for gravitational waves, and spectroscopy of atoms.

Superconductors

At low temperatures, some materials conduct electrical current without any loss. Below a certain "transition" temperature, electrical resistance disappears entirely. This is a quantum-mechanical effect. For most superconductors such as niobium and lead, or niobium-titanium alloy, this transition temperature lies a few degrees above absolute zero, or minus 273 degrees, so that liquid helium has to be used as a coolant. A few newly discovered ceramic materials, so-called perovskites, become superconductive at higher temperatures so that cooling with liquid nitrogen suffices. Quantum mechanics offers an explanation for this jump in electrical resistivity, and indicates in what direction one has to look to find new superconductors.

The quantum mechanical explanation describes superconductivity as a result of the fact that pairs of electrons move through the crystal lattice without resistance, as the theoreticians Bardeen, Cooper, and Schrieffer discovered. In contrast to the normal electrical conductivity, which is brought about by the movement of electrons in the "conduction band" of the metal, here it is two electrons with opposite "spin" that make up a pair. Each of these pairs has a combined spin zero; they obey the Bose-Einstein statistics, and can therefore be in the same quantum state and move synchronously, like the photons in a laser.

Superconductive magnetic coils for the generation of extremely high field forces are required in medical magnetic resonance imaging and in high-energy particle accelerators. Because of the high energy loss through heat buildup, conventional magnets are unusable in these applications.

Magnetic Resonance Imaging

The constituent elements of the atomic nucleus are the charged protons and the uncharged neutrons. Like electrons in the atomic shell, both have a "spin." The proton has in addition a magnetic dipole moment. It acts as an elementary magnet, like the electron. Its strength is about a thousand times smaller than that of the electron but nevertheless can be detected.

The elementary magnets in an external magnetic field align themselves along the field, either in the direction of the field or opposite to it. The magnetic moment can be turned over by a high-frequency electrical alternating field with a certain frequency. If the same amount of energy is supplied to the atom by the electrical field as is required for the folding of the magnetic moment of the proton, an electrical signal is obtained from which the position of the proton can be determined. This method is called nuclear magnetic resonance, or NMR. It can be applied in magnetic resonance imaging (MRI) to obtain very precise images of soft tissues containing water – i.e., the nucleus of hydrogen protons. By contrast with X-ray pictures, bones are barely visible because they contain little water, whereas brain matter and cartilage in joints are excellently imaged.

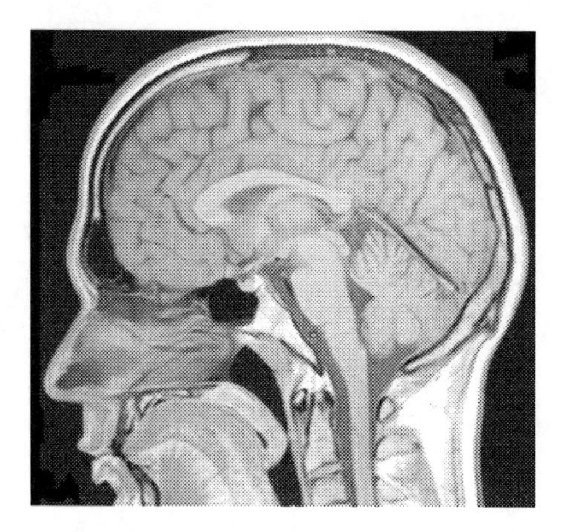

Fig. 4.10 One of many applications: the MRI process in medicine, a picture of a brain

5

Expulsion and the War Years

Einstein and Germany

Einstein's relationship to Germany went through highs and lows. In 1954, one year before his death, he wrote a brief autobiography beginning with the words, "I was born a German in Ulm in 1879. I spent my youth in Munich, where I attended high school. In 1895, following a brief stay in Italy, I went to Switzerland. From 1896 to 1900, I studied mathematics and physics in Zürich at the Swiss Polytechnic."

Seven years before Einstein's birth, the kingdom of Württemberg had become a member of the German Reich, with the special privilege that its inhabitants retained Württemberg citizenship. One of the reasons Einstein left the high school in Munich bound for Milan was to avoid the looming draft into military service. Because his application for admission to study at the Swiss Polytechnic in Zürich without a high school diploma was rejected, he had to make up the high school certificate in Aarau. At this time he applied for release from Württemberg citizenship, which was granted in January, 1896.

For the next five years he was technically stateless. Following his course of study, he applied for Swiss citizenship. On February 21, 1901, he was granted Zürich citizenship and thereby became Swiss.

Having traveled the path from Bern to Zürich to Prague and back once more to Zürich, Einstein finally became a full and permanent member of the Prussian Academy of Sciences in 1913, and thereby *de jure* a Prussian and again a German citizen. During the war, while he was working intensively on the general theory of relativity, achieving his decisive advances, his health

K. Kleinknecht, *Einstein and Heisenberg*, Springer Biographies,
https://doi.org/10.1007/978-3-030-05264-5_5

declined. Separation from Mileva, a bachelor's irregular life style, inadequate food, not much mitigated by occasional packages from Switzerland, all took its toll on his strength.

After Germany's defeat in 1918, Einstein's Zürich friends, Professor Heinrich Zangger in particular, saw the opportunity to win him back for Zürich. In 1918, they succeeded in having an offer extended to Einstein of a joint professorship at the ETH and the University of Zürich. He declined. He explained his refusal in a letter to Michele Besso:

> *But if you could see how wonderful relationships among my closest colleagues (especially Planck) and me have developed, and how accommodating everyone here has been and continues to be, and if you further bear in mind that my work has come to fruition only through the understanding it has found here, then you will understand that I cannot decide to turn my back on this place.*

Planck and Nernst had been the first to take Einstein seriously as a scientist and recognize the importance of his papers, as well as helping him be recognized in the scientific community. He did not forget this. Already earlier, on February 8, 1918, he had written in a letter to Hedwig Born: "To live next to Planck is a pleasure." He met with his close colleagues regularly, as, for instance, at the time of the Berlin visit of the American Nobel laureate Robert Andrews Millikan in 1931.

After the abdication of Kaiser Wilhelm II on November 9, 1918, and the revolution of the Workers' and Soldiers' Councils, Einstein was enthusiastic about the end of the Kaiser Reich and the path to the Weimar Constitution. He wrote his mother, "Up to now it has all gone smoothly, even impressively. I am very happy about the development of things. For the first time, I'm feeling good here." Together with Max Born and Max Wertheimer, he rode to the Parliament building to follow the deliberations. He pleaded to the student council for a parliamentary democracy and against the Soviet system.

Fig. 5.1 Einstein with Nernst, Planck, Millikan, and v. Laue, 1931

When Arnold Sommerfeld wrote him, "I hear that you believe in the new time, and want to contribute to it," he replied: "It is true that I have hopes for this time. I am of the firm conviction that culturally aware Germans will soon once more be able to be as proud of their fatherland as ever – with more reason than before 1914."

The spectacular confirmation of the general theory of relativity by the observation of light deflection in the gravitational field of the Sun by the expedition of the Royal Astronomical Society naturally became a political event, too. The theory of gravitation by a German had toppled Newton from his throne, as one newspaper put it. In 1919, he wrote to the *London Times*: "Today I am called a 'German scholar' in Germany; in England, a 'Swiss Jew.' Should I ever be in the position to be presented as a *bête noire*, however, I would conversely be a 'Swiss Jew' to the Germans, and a 'German scholar' to the English."

In November 1922, the Nobel Committee awarded Einstein the prize for physics, although he was at the time traveling through Japan and unable to attend the awards ceremony. A competition between the German and Swiss ambassadors ensued over who should accept the prize in his name. The Prussian Academy announced that with his acceptance of a position at the Academy, he had become German *de jure*, and the German ambassador received the prize.

Fig. 5.2 Homeward bound from Japan with Elsa, 1922

On his return in 1922, Einstein questioned the wisdom of his having taken on German citizenship, but he ultimately accepted the fact. Now he was a dual citizen. In 1925, the Swiss embassy in Berlin declined his application for a diplomatic passport to travel to Geneva as member of the International Commission on Intellectual Cooperation, as well as to South America. Instead he did receive a German diplomatic passport, because the foreign office always set store by Einstein's traveling with a German passport.

The ascendant National Socialism and anti-Semitism turned Einstein into an object for their slogans of hate. While Max von Laue, Max Planck, and Werner Heisenberg defended relativity theory in public lectures, and regarded it as a great step forward and an indispensable element of modern physics, there was a contingent of professors who rejected relativity theory as "Jewish," and polemicized against it. In particular, it was Philipp Lenard in Heidelberg and Johannes Stark, president of the Physical-Technical Reich Institute in Berlin, who sought to establish a "German physics."

Einstein's optimism with regard to the progress of the Weimar Republic had vanished. On July 17, 1931, he wrote Max Planck that he intended to relinquish his German citizenship, while retaining his position at the Academy of Sciences. The importance of Einstein, Planck, von Laue, Nernst, and Haber notwithstanding, Berlin was no longer the center of physics; Göttingen and

Leipzig had grown more important through the development of quantum mechanics by Heisenberg, Born and Jordan.

In September 1932, Einstein went even further. He informed the Prussian ministry that he had committed to be in Princeton for the winter semester of 1932-33. "These commitments are naturally incompatible….with the requirements of my position as member of the Prussian Academy of Sciences. The question therefore is whether maintaining my position at the Academy under the new circumstances is even possible or desirable."

In December 1932, Einstein traveled with Elsa to America. In January 1933, Reich President Hindenburg appointed Hitler Reich Chancellor, thus enabling the seizure of power by the NSDAP and the transformation of the democracy into a dictatorship. On March 23, 1933, the parliament enacted the "Enabling Act." Einstein reacted resolutely. He wrote a strong declaration against the acts of brutal violence and repression, and offered it to the *International League against Racism and Antisemitism.*

On February 28, 1933, even during the return trip on the ship, he composed his declaration of withdrawal from the Academy. It was read out at a session of the Academy on March 30, 1933. The Academy had given him the opportunity for 19 years to dedicate himself free of all professional duties to scientific work, but, "I find my dependence on the Prussian government that is contingent on my position unacceptable under the present circumstances."

As a closure to his time in Berlin, on April 4, 1933, he applied from Oostende for release from Prussian (i.e., German) citizenship. The reaction in Germany demonstrated that most people had not yet recognized how dangerous the anti-Semitism of the National Socialist regime for Jewish citizens was. Even German Jews criticized Einstein.

After his stay in Belgium, Einstein returned to America. In October 1933, he arrived in New York and settled in Princeton. Two years later, he moved into his house at 112 Mercer Street. He declared he would never return to Germany.

It is possible that his declaration notwithstanding, Einstein did visit Germany in June 1952, as a letter of July 20 of the same year sent to the director of the Castle Museum, Büdingen (Hesse) suggests.

Twenty-two-years old at the time, Rainer Lott, a physics student was a contemporary witness with whom the author spoke in September 2015 in Murnau. At the time, he had been in his first semester at the University of Giessen and returned home that evening to Büdingen. There he ran into a close friend, Erhart Karrer, two years younger, who was still attending high school. Erhart told him he had that day taken his German teacher and Einstein through the medieval Büdingen. Naturally, Rainer Lott was impressed. When his father, Friedrich Karl Lott, came home, he related to his son that on his way to his elevated stand in the woods that day, on the square between the Jerusalem Gate and the *Stern* restaurant, he had run into two gentlemen – one, his son's teacher, Dr. Josef Neupärtl, the other also a professor. Dr. Neupärtl had introduced him by name. The guest had asked after his profession. On learning he was a geometer at the Bureau of Weights and Measures, Einstein replied, "Well then, we have nearly the same profession."

The two men also visited the Castle Museum. On July 5, the director of the museum, who had not been present the day of the visit, wrote to Einstein in Princeton. He regretted not having been there at the time of his visit. By way of broadening his impressions, he was sending along a guidebook of the museum. On July 20, 1952, Einstein thanked him for both letter and guidebook. "Your kind letter and the tasteful pamphlet have recalled for me my visit to your Arcadia. A bit of the Middle Ages shown from its most attractive side. My sincere thanks for this kindness. With great respect, Albert Einstein."

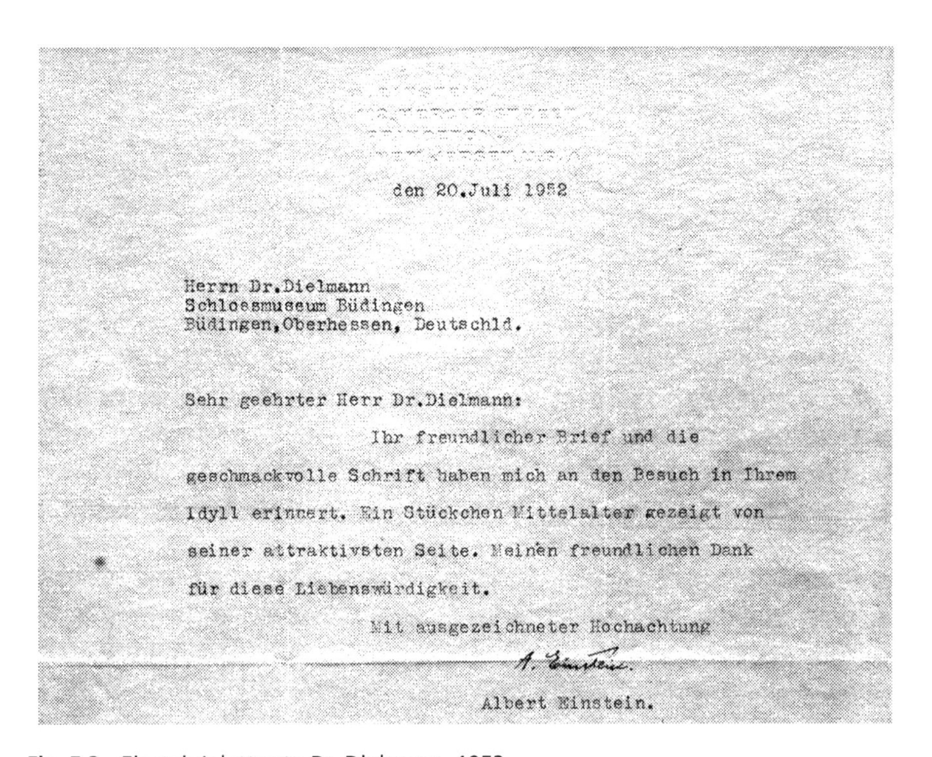

den 20.Juli 1952

Herrn Dr.Dielmann
Schlossmuseum Büdingen
Büdingen,Oberhessen, Deutschld.

Sehr geehrter Herr Dr.Dielmann:

Ihr freundlicher Brief und die geschmackvolle Schrift haben mich an den Besuch in Ihrem Idyll erinnert. Ein Stückchen Mittelalter gezeigt von seiner attraktivsten Seite. Meinen freundlichen Dank für diese Liebenswürdigkeit.

Mit ausgezeichneter Hochachtung

A. Einstein.

Albert Einstein.

Fig. 5.3 Einstein's letter to Dr. Dielmann, 1952

Einstein's Pacifism, the Bomb, and the Franck Report

From the time of his youth, Einstein had a strong aversion to being compelled to do anything he didn't want, be it in school or adhering to any scientific or religious doctrine he held to be false. He especially hated the military with its command structure and its requirement for obedience. Thus, it was consistent with his personality that he dodged the draft in Munich by his flight to Italy. He was fortunate not to have to complete the obligatory military service following his adoption of Swiss citizenship because he was rejected as unfit to serve. Thus, his positive view of Switzerland was kept intact.

After accepting the invitation to Berlin despite reservations about Prussian militarism, he did not demonstrate any political support for the war but rather dedicated himself to working out his general theory of relativity. When the war had been lost and the Kaiser abdicated, he saw his big chance. "The great thing has happened," he wrote his sister. "That I should live to see this!.... Militarism and privy counselor fog are thoroughly eliminated." And to his mother: "Now I am really enjoying myself here. The collapse has done wonders. Among the academics, I'm a kind of upper socialist." Politically, he was active with the social democrats. He was also sympathetic towards the communists in Russia and Germany.

Since 1915, Einstein had been a member of the New Fatherland League, which later renamed itself the German League for Human Rights. In telegrams to the Reich chancellery and the ministers in 1918, the league demanded freedom for political prisoners, freedom of assembly, speech, and press, equal and secret voting rights for a national assembly, and the "eradication of human need by socialization of the means of production," in short the establishment of a socialist society. The party program was worked out and signed by an executive committee, whose members included Magnus Hirschfeld, Albert Einstein, Max Wertheimer, Heinrich Mann, Graf Arco and Käthe Kollwitz. In June 1924, Einstein requested a meeting with Reich chancellor Dr. Marx, which was granted just five days later. He advocated for the freeing of one of the chief communist functionaries in the Bavarian soviet government, Erich Mühsam, who had been sentenced to fifteen years in prison. One year later, Mühsam was granted early freedom.

Another membership was added in 1923: the Society of Friends of the New Russia, whose central committee Einstein joined. The society's literature said: "Germany and Russia, which are economically and intellectually connected, have the greatest interest in coming together.... To eliminate the disastrous effects of the world war, the cultural workers of both countries must join together in a cultural community."

In an annotation, the foreign office recorded that the political orientation of the society was to be understood in the sense of an affirmative attitude towards communism.

Einstein's membership on the board of the Red Help ("Rote Hilfe"), a relief organization of the German communist party, can be seen in the same light. The Red Help was a division of the Workers' International Relief (WIR), whose German branch was established in 1925 by the German communist party. The first chairmen were Wilhelm Pieck and Clara Zetkin. In 1927, Einstein was actually elected to the Central Committee of the WIR. "Politics is actually evolving consistent with the Bolshevik side," he wrote on January 17, 1920, to Max Born. He was very complimentary to the Polish-German communist Karl Radek, who had been sent from Moscow to participate at the founding congress of the German communist party. He expressed praise even for Lenin: "In Lenin, I honor a man who has self-sacrificingly applied his whole strength for the realization of social justice."

When a group of European intellectuals in a declaration criticized a political trial in the Soviet Union against "the 48 miscreants," he signed at first but regretted his signature later because he considered criticism of the Soviet Union not right. Before the parliamentary election of 1932, in an open letter he, along with Heinrich Mann and Käthe Kollwitz, demanded that the chairs of the German social democratic party and the German communist party, Otto Wels and Ernst Thälmann, and the chair of the trade union ADGB, Theodor Leipart, form an anti-fascist unity front and set up common lists of candidates, without success.

Fig. 5.4 Poster of the appeal to the SPD and KPD

Many years later, in 1944, he wrote Max Born that he had to laugh bitterly today at the memory of 1918, that he had once thought it possible to make honest democrats of those characters. As men of 40, they had both been naïve.

Aside from his work in political organizations, Einstein also participated increasingly in Zionist groups. He was a member of the presidium of the Jewish Peace League, and his wife Elsa participated in the Women's Committee of the Peace League. In a message to the Jewish community, he appealed for a "voluntary Jewish peace tax" to be instituted to assure the participation of Judaism in the work of peace. His trips to America in 1921 and 1931 were promotions for the establishment of a Jewish state in Palestine, and for a strict pacifism. His appearance was sometimes accompanied by a mass hysteria. Public interest and press coverage were overwhelming. At banquets for the Palestine campaign, at which Einstein was the guest speaker, considerable sums were donated for the colonization of Palestine.

During this time, Einstein supported a "militant pacifism." He called for refusal of military service, "regardless of one's judgment of the cause of a war." Even if a nation is invaded, it had no right to defend itself. However, this changed the moment the National Socialists seized power in Germany.

Now he declared:

> *Because Germany is obviously preparing with all its resources for a war, France and Belgium find themselves in grave danger, and are unconditionally reliant on their armies.... Under the current circumstances, were I a Belgian, I would not resist military service but rather would gladly take it upon myself with the feeling of rescuing European civilization.*

His pacifist-minded friends were horrified that at the moment his pacifism encountered reality, he abandoned his principles. Romain Rolland wrote on September 15, 1933, to Stefan Zweig:

> *Einstein is more dangerous to a cause as a friend, than as your enemy. He is a genius only in his science. In other areas, he is a fool.... His declarations two years ago on refusal of military service in America were absurd and untenable.... To make young people believe their resistance could stop the war amounts to criminal naiveté.... Now he does an about-face, and condemns the conscientious objectors with the same nonchalance with which yesterday he supported them.... He was created only for his equations.*

The looming world war and Otto Hahn's and Fritz Straßmann's discovery of the splitting, or fission, of the heavy element uranium, as well as the calculation by Lise Meitner and Otto Frisch of the enormous energy released in this fission process, alarmed physicists. The Hungarian-born physicist Leo Szilard anticipated the possibility of using such a process for a chain reaction and in secret had obtained a British patent for it much earlier, in 1934. Immediately after the discovery in Germany, Enrico Fermi repeated the experiments in Chicago and confirmed the results.

Together with Szilard, Fermi attempted to produce a chain reaction in a small reactor. In the fission of uranium, the neutrons released had to be slowed down in order to produce further fission. Water was unsuitable for slowing the neutrons since it absorbed the neutrons. Szilard, who had studied chemistry and worked from 1919 to 1933 in Berlin, hit on the idea of using highly purified graphite. Care had to be taken that the graphite contained no impurities that might contain the neutron-absorbing element boron. However, there was not enough money to procure uranium and graphite in large quantities. Szilard and his colleague Eugene Wigner were convinced that the United States should develop a weapon against Germany from this. To have a chance to get to the American president with this proposal, Szilard planned to win over the best known, most prominent physicist in the world to write a letter to the president. So on July 12, 1939, Szilard and Wigner visited Einstein on Long Island. Szilard had become acquainted with Einstein in Berlin, where they had worked together on a patent for a refrigerator.

When the visitors came to Einstein´s summer resort, Einstein told them he had not even considered the possibility of building a bomb with uranium, but he agreed to use his authority for such a letter and to take responsibility for it. So Szilard drafted a letter and visited Einstein for a second time on August 2, 1939. This time the driver was a former doctoral student of Heisenberg's, Eduard Teller. It was a peaceful summer day, no war anywhere.

Einstein was prepared to write the American president, although previously he had always held pacifistic convictions and rejected weapons production. The fear was that the Germans might build such a bomb. His only criticism was that the draft letter was too long and in part incomprehensible. He wanted a shorter version with a clear message. He dictated his own version, a short German sketch. Over the next few days, Szilard translated this text, incorporating edits. Finally, two versions resulted, a long and a short version, which Szilard sent to Einstein in August, 1939. Einstein returned both versions with his signature, but preferred the long one.

Szilard learned from a friend, the Austrian journalist and former member of German Parliament Gustav Stolper, that the best chance for having the president actually read the letter was a personal delivery by a person with access to the president. Alexander Sachs was such a person whom Stolper knew, and who was prepared to act on the matter. Sachs was an investment banker at Lehman Brothers. In 1932, he had written the economic portions of Roosevelt's campaign speeches and was known to the president through his activity on various administration committees, most recently since 1936 the National Policy Committee. Szilard gave the final version of the letter to Sachs to deliver personally to the president.

It took nearly ten weeks, though, before Sachs got the opportunity to deliver Einstein's letter to Roosevelt. In the meantime, on September 1, 1939, Germany invaded Poland. In his book *Brighter Than a Thousand Suns*, Robert Jungk tells the story of this dramatic encounter on October 11, 1939. Roosevelt sat in a wheelchair in his office in the White House and received

Sachs. To make sure the president would acknowledge the contents of the letter, instead of putting it aside and forgetting it under the pile of other documents requiring his attention, Sachs read him the letter aloud, along with an additional memo from Szilard and a comprehensive letter of his own.

The effect of the hour-long reading was not as overwhelming as Sachs had expected and hoped. Roosevelt was exhausted by the lengthy recitation and saw no reason for taking the matter on. He told the disappointed Sachs that he thought it was interesting but considered government intervention to be premature at this moment.

At their leave-taking, Sachs succeeded in obtaining an invitation of the president to breakfast the next morning. Sachs did not sleep that night. He deliberated feverishly how he could persuade Roosevelt of the necessity of taking action. Three or four times during the night he left the hotel to meditate in a nearby park: "What could I say to get the president on our side in this affair, which was already beginning to look practically hopeless? Quite suddenly, like an inspiration, the right idea came to me. I returned to the hotel, took a shower and shortly afterwards called once more at the White House."

Roosevelt was sitting alone at breakfast. At his entering, he asked Sachs ironically, "Alex, what bright idea have you got now? How much time would you like to explain it?" Sachs replied that he would not take long:

> *All I want to do is to tell you a story. During the Napoleonic wars a young American inventor, Robert Fulton, came to the French Emperor and offered to build a fleet of steamships with the help of which Napoleon could, in spite of the uncertain weather, land in England. Ships without sails? Nonsense! This seemed to Napoleon so impossible that he sent Fulton away. In the opinion of the English historian Lord Acton, this is an example of how England was saved by the shortsightedness of an adversary. Had Napoleon shown more imagination at that time, the history of the nineteenth century would have taken a very different course.*

When Sachs had finished, the president remained quiet several minutes. Then he wrote a few words down on a slip of paper and gave it to his assistant, who waited at the table. After a while, the assistant returned with a package which, on Roosevelt's instructions, he carefully unpacked. It was a bottle of V.S.O.P French Napoleon cognac, which the Roosevelt family had had for many years in their cellar. Silently, he signaled the assistant to fill two glasses, lifted his, and toasted Sachs. And then he observed, dryly, "Alex, what you are after is to see that the Nazis don't blow us up?"

"Precisely."

Then Roosevelt called in his attaché, General Edwin "Pa" Watson, and told him –referring to Einstein's letter and Szilard's and Sachs' documents – a sentence which later became famous: "Pa, this requires action!"

The same day, Watson called in Dr. Lyman Briggs, director of the National Bureau of Standards, and discussed with him the formation of a committee with representatives of the army and the marines, as well as physicists to deliberate over the uranium question. The president at once informed Sachs of the decision, who left satisfied. In this way, the program for building the bomb had arrived at the highest level.

The letter of Einstein to President Roosevelt said the following:

Albert Einstein
Old Grove Road
Peconic, Long Island
August 2nd, 1939

F. D. Roosevelt
President of the United States
White House
Washington, D.C.

Sir:

Some recent work by E. Fermi and L. Szilard, which has been communicated to me in manuscript, leads me to expect that the element uranium may be turned into a new and important source of energy in the immediate future. Certain aspects of the situation which have arisen seem to call for watchfulness and if necessary, quick action on the part of the Administration. I believe therefore that it is my duty to bring to your attention the following facts and recommendations.

In the course of the last four months it has been made probable through the work of Joliot in France as well as Fermi and Szilard in America – that it may be possible to set up a nuclear chain reaction in a large mass of uranium, by which vast amounts of power and large quantities of new radium-like elements would be generated. Now it appears almost certain that this could be achieved in the immediate future.

This new phenomenon would also lead to the construction of bombs, and it is conceivable – though much less certain – that extremely powerful bombs of this type may thus be constructed. A single bomb of this type, carried by boat and exploded in a port, might very well destroy the whole port together with some of the surrounding territory. However, such bombs might very well prove too heavy for transportation by air.

The United States has only very poor ores of uranium in moderate quantities. There is some good ore in Canada and former Czechoslovakia, while the most important source of uranium is in the Belgian Congo.

In view of this situation you may think it desirable to have some permanent contact maintained between the Administration and the group of physicists working on chain reactions in America. One possible way of achieving this might be for you to entrust the task with a person who has your confidence and who could perhaps serve in an unofficial capacity. His task might comprise the following:

a) to approach Government Departments, keep them informed of the further development, and put forward recommendations for Government action, giving particular attention to the problem of securing a supply of uranium ore for the United States.

b) to speed up the experimental work, which is at present being carried on within the limits of the budgets of University laboratories, by providing funds, if such funds be required, through his contacts with private persons who are willing to make contributions for this cause, and perhaps also by obtaining co-operation of industrial laboratories which have necessary equipment.

I understand that Germany has actually stopped the sale of uranium from the Czechoslovakian mines which she has taken over. That she should have taken such early action might perhaps be understood on the ground that the son of the German Under-Secretary of State, von Weizsäcker, is attached to the Kaiser-Wilhelm Institute in Berlin, where some of the American work on uranium is now being repeated.

Yours very truly,
Albert Einstein

On October 19, 1939, Roosevelt thanked "[his] dear Professor" Einstein for his letter and its very interesting and important contents. He had taken the contents so seriously that he immediately assembled a committee, comprised of the president of the National Bureau of Standards and representatives of the army and the marines, to deal intensively with the possibilities of Einstein's proposal regarding the element uranium. The president formed a uranium committee.

The committee met for the first time on October 21, with nine participants: Briggs and his assistant, Sachs, Szilard, Wigner, Teller, Roberts, Adamson (from the army), and Hoover (from the navy). Szilard spoke at once of a bomb that could have the explosive power of 20,000 tons of TNT. Other members of the committee were skeptical, but when Hoover asked how much money the physicists needed, Teller cited a sum of $6,000 for the acquisition of high-grade graphite for the moderator for Fermi's reactor. This sum was much too little, as Szilard wrote Briggs a few days later. The graphite alone would cost $33,000.

In the committee's first report to the president, first and foremost the application of a small reactor to drive submarines is cited. Whether a bomb of great destructive power was possible would be clarified through fundamental investigations, for which appropriate funding was recommended.

When not much happened in the months following his first letter, Einstein wrote Roosevelt a second letter on March 7, 1940, which was again to be delivered by Sachs. In this letter he mentioned that Szilard had shown him a manuscript about a chain reaction in uranium, and raised the question of whether publication of this paper should be prevented. On March 15, Sachs delivered Einstein's second letter to Roosevelt, and on April 5, the president arranged for the uranium committee to be enlarged.

We can date the beginning of the Manhattan Project for building the bomb to Saturday, December 6, 1941. Vannevar Bush, director of the Bureau of Research and Development, and James B. Conant, chairman of the National Defense Research Committee, called the uranium committee together to reorganize the work. Enrichment of the bomb material, U-235, was now to be undertaken with two independent methods, gas diffusion and electromagnetic separation. The start-signal for the building of the bomb, then, was given at a time when America had not yet entered the war.

One day later, Japan attacked the American fleet in Pearl Harbor. The impetus for the uranium project was increased. The definitive breakthrough came with the first evidence of a self-contained chain reaction with natural uranium and graphite in the experimental reactor that Enrico Fermi set in operation in Chicago in December 1942. With uranium and graphite purchased with funds provided by the committee, Fermi and Szilard succeeded in 1942 in Chicago in triggering a controlled chain reaction with slow neutrons. The test reactor consisted of 5.4 tons of natural uranium, 45 tons of uranium oxide and 360 tons of purified graphite and had a volume of 164 cubic meters. Fermi and Szilard were granted a U. S. patent for such a reactor in December 1944.

The rest of the story is well known. The basis of the American bomb was the separation of the fissionable isotope U-235, which exists only as 0.7% of the natural uranium, and the production and chemical separation of the similarly fissionable plutonium in large power reactors. Both processes were carried out at enormous expense in order to produce the critical mass of fissionable material necessary for a bomb.

The Manhattan Project employed up to 150,000 workers, engineers, and researchers, and cost $2 billion at that time, which corresponds to about $27 billion today. In Oak Ridge, Tennessee, a huge plant was built with an area of 170,000 square meters for the separation of isotopes by gas diffusion of uranium-hexafluoride, as well as subsequent further enrichment of the U-235 in an electromagnetic mass spectrometer. In Hanford, Washington, several power reactors for the generation of plutonium were built and a factory for separating plutonium from the fuel rods. In Chicago a metallurgical laboratory was constructed, and in Los Alamos, New Mexico, there was a secret laboratory for construction of the bomb, with Robert Oppenheimer as scientific director and General Leslie Groves as military commander. The scientists were consistently told they were in competition with a German project, although the British Secret Service (and thereby also the Americans) knew clearly from the reports of the spy Paul Rosbaud (code-named the griffon) how insignificant the activities of the German "Uranium Club" were in comparison. By the end of the war, the German physicists had not even succeeded in triggering a chain reaction in a test reactor, a step Enrico Fermi and Leo Szilard had achieved by 1942 in Chicago.

Fig. 5.5 Factory for the separation of isotopes, Oak Ridge, TN

In 1940, Einstein acquired American citizenship; on April 12, 1945, on the death of Roosevelt, Harry S. Truman became president of the United States. On May 8, Germany surrendered unconditionally.

In this moment when the goal of the war had been partially realized, a group of scientists from the University of Chicago who had taken part in the Manhattan Project for the development of the atomic bomb organized themselves as the Committee on Political and Social Problems. James Franck, former colleague of Max Born in Göttingen, initiated the group. He, like most of the German immigrants on the Manhattan Project, worked on the creation of the bomb after he had acquired U. S. citizenship. As director of the chemical division of the metallurgical laboratory at the University of Chicago, he began with the extraction of plutonium for the building of the atomic bomb in 1942. Following the German surrender, though, he had moral misgivings about the use of atomic weapons, in particular against the now planned dropping of the bomb on a Japanese city. So he established the committee, and together with other members worked out a comprehensive rationale against dropping the bomb. He argued that in dropping it on a city, the United States would "lose the public support of the world," and "call forth an arms race." Instead, he proposed demonstrating the "new weapon" to the representatives of all lands of the United Nations on uninhabited territory.

To Franck's great disappointment, many of the scientists at Los Alamos, the other major laboratory of the Manhattan Project, did not sign the petition. Neither Fermi, Oppenheimer, Bethe, Feynman, Teller, nor Weisskopf signed the letter. General Groves had prohibited circulation of the petition inside the Los Alamos laboratory by an administrative trick. Also the signature of the pacifist Einstein is missing on the document. His great political weight might perhaps have had some effect. On the other hand, a total of 70 scientists did sign.

On June 12, 1945, Franck personally delivered the resolution, which went down in the history books as the Franck Report, to George L. Harrison, Special Assistant to the U. S. Secretary of War Henry L. Stimson in Washington, to prevent in the last minute the catastrophe of an explosion over a Japanese city. This was barely two months before the dropping of the atomic bomb on Hiroshima on August 6.

Harrison was a member of the secret Interim Committee, which President Truman had established in May 1945 to advise him on the question of nuclear energy. On June 21, the committee met and debated the use of the atomic bombs, and confirmed its earlier recommendation to use the bombs as soon as possible and without warning, to inflict the greatest possible damage to the military installations and residential buildings of the target city. The appeal of the Franck Report was overruled by the opinions of four scientists from the Manhattan Project who had been called in – Enrico Fermi, Arthur Compton, Ernest Lawrence, and Robert Oppenheimer – who endorsed the drop and who enthusiastically celebrated their success in Los Alamos after the destruction of Hiroshima.

On the afternoon of August 6, 1945, radios in America reported the dropping of a new kind of bomb on the Japanese city of Hiroshima. Helen Dukas, Einstein's secretary, heard the news and recalled "the Szilard thing" he had discussed with Einstein in 1939. When Einstein got up from his noontime nap, she told him what she had heard. Einstein said only, "Oh, Weh!" When

Szilard visited him shortly thereafter, they came to discuss the events six years before on Long Island when they had discussed the letter Einstein was to write to Roosevelt. Einstein said, "There you see the ancient Chinese were right. It is impossible to predict the consequences of one's own actions. All one can do may be to do nothing."

Subsequently, he lamented his actions promoting the atomic bomb: "I made a great mistake when I signed the letter to President Roosevelt in which I advised production of the atomic bomb. Had I known the Germans would be unsuccessful with it, I would not have lifted a finger."

And on December 20, 1948, he wrote to Carl Gustav Jung, "I have always spoken out against violence, but my theories have unfortunately put the potential for the most terrible violence into the hands of mankind, and this is a heavy burden for me."

It was not just the theories that made the bomb possible, though, but also his letters to President Roosevelt.

When the British philosopher Bertrand Russell later issued a call against a nuclear arms race, Einstein signed the petition on April 11, 1955, one week before his death. Eight other eminent scientists, among them Max Born and the chemist Joseph Rotblat, who had worked on the Manhattan Project and withdrawn from it in 1944, signed the appeal, which was published on July 9, 1955. The group established the first Pugwash conference against the nuclear arms race in 1957.

Heisenberg, the War Years, and the Uranium Club

Political developments following Hitler's annexation of large parts of Czechoslovakia intensified fears that a war threatened. For this reason, in the spring of 1939, Heisenberg sought out a country house in the mountains for his family, in which his wife and children could find refuge if the cities were destroyed. Wilhelmine, the daughter of the painter Lovis Corinth, offered them a wooden house her father had built for the summer vacation months. It was located in Urfeld above Lake Walchen on the south-facing slope, about hundred meters above that road on which the students Wolfgang Pauli, Otto Laporte, and Werner Heisenberg had once done a cycling tour, discussing quantum theory. It became the family's getaway house.

In view of the precarious political situation and a looming war, Heisenberg traveled to America in June 1939. He had many friends there, and felt the need to see them once more before war broke out. He wrote in his autobiography: "One didn't know if one would ever meet again. Should I be able to work on the reconstruction after the catastrophe, I hoped for their help, too,"

So he traveled on the *Mauretania* from Liverpool to New York, met with old friends in Chicago, and then delivered a series of lectures over the course of several weeks at Purdue University. There were ten times as many people in his audience as in Leipzig, and he was treated splendidly in every respect. Ultimately, the head of the department made him a concrete job offer. Heisenberg wrote his wife:

> *The light- and shadow sides are both so terribly apparent. Right away, I would have ten times as many talented students as at home. No doubt more would come of my work, too. But we simply are not at home here. The children would speak English, and grow up in a way that is foreign to us. Just for this reason we'll stay at home.*

But he saw other advantages in America. When the master craftsman invited the entire institute and was regarded as on equal standing with the professors, Heisenberg observed that therein lay the actual strength of America: that class differences did not exist here. Nevertheless, he declined the offer, as he did those from other American universities.

On July 30, 1939, in New York he bade "farewell to the world of man-made peaks to travel to the friendlier mountains formed by nature," namely close to the Urfeld house, which had been furnished in the meantime, where he spent August with his family. From there, he planned to take part in a conference in Zürich, and wrote on August 24 to his friend Pauli to please send him the program of the session, and to arrange for the honorarium promised him by Paul Scherrer to be deposited with the hotel City Excelsior, because he would be arriving in Zürich without money. Obviously, Heisenberg was not expecting that the war, the looming "catastrophe," would come so soon.

The session was not to take place. On September 1, 1939, the Second World War broke out as German troops invaded Poland. Heisenberg expected to be drafted into the Mountain Troops, whose maneuvers he had taken part in three years earlier. But things turned out differently. Apparently, there was a dispute between two departments over his assignment. On September 15, he heard from Peter Debye from Berlin that he would be assigned to an Army Ordnance Department or something of the kind. On September 20, he was conscripted to Berlin to a scientific position, but then sent back to Leipzig.

In these weeks Kurt Diebner, an experimental physicist employed by the Army Ordnance Department, summoned a group of leading nuclear physicists to the agency to explore the possibilities of applying the fission of the uranium nucleus, discovered in 1938 by Otto Hahn and Fritz Strassmann, to the building of an energy-generating machine or a bomb. Present in addition to Diebner was his superior, Dr. Basche, and a group of physicists who later dubbed themselves the Uranium Club. Included were Paul Harteck from the University of Hamburg, Erich Bagge from the Army Ordnance Department, and Karl Wirtz from the Kaiser-Wilhelm Institute for Physics in Berlin. Heisenberg was not invited. When Harteck was asked by Basche if he would take over leadership of the entire project, he declined. He argued that such a project was of an order of magnitude that small university groups of one professor each could not control. Something like this required the collaboration of technicians, engineers, experimental physicists, and the enlisting of industry. In the discussion of who else might be invited to join the group, Harteck named Gustav Hertz as the best of the experimental physicists.

In 1935, Hertz's certification to examine doctoral candidates at the Technical University, Charlottenburg, had been revoked on the grounds of his Jewish ancestry; nevertheless he acted as doctoral supervisor still in 1938. He resigned his position at the Technical University and became leader of the research division of Siemens and Halske in Berlin. He had developed a process for enriching isotopes of noble gases that could be utilized to separate the fissionable isotope of uranium, U-235. The uranium had to be dissolved in hydrofluoric acid and converted into the gas form uranium hexafluoride.

This process would later be applied in the American Manhattan Project in the huge plant at Oak Ridge, and after 1945 in the soviet bomb project. But Harteck's suggestion to assign Hertz a leading role in the project was rejected. Instead, Bagge proposed his doctoral advisor, Heisenberg, who albeit as a theoretician was not ideally suited to develop a new technology, had an outstanding reputation as a Nobel laureate. Heisenberg had never carried out an experiment; he had neither instinct for experimental and organizational details, nor any interest in them. His "real" work had always been theoretical analysis and research. He was the wrong man for the project.

Conscription to the Army Ordnance Department in Berlin followed several weeks later. In this way, he was spared military service; he was "indispensable." It was disclosed to him in Berlin that, together with other leading nuclear physicists, he was to research and find applications for the fission of uranium. The Kaiser-Wilhelm Institute for Physics (KWI) in Berlin-Dahlem was put under the control of the Army Ordnance Agency as the center for uranium research. The Director of this institute, the Dutchman Peter Debye was given the choice of adopting German citizenship or taking a leave of absence. In January 1940, he emigrated to the United States and was Guest Professor of Chemistry at Cornell. Kurt Diebner, an experimental physicist from the Army Ordnance Department, was assigned to represent and replace him as leader of the Berlin uranium project.

The KWI was a center of uranium research, to be sure, but the activities of the Uranium Club were fragmented. Small groups, each comprised of a professor or group leader and a few assistants, worked at five different locations: the institutes in Leipzig, Dahlem, Hamburg, Heidelberg, and in the army's research site Gottow near Kummersdorf in Brandenburg.

Heisenberg was to function as scientific advisor at the Berlin Institute. So he traveled at the start of each week from Leipzig to Berlin, delivered his lectures in Leipzig in the latter part of the week, and did theoretical work on the uranium project.

In the fission of the uranium nucleus with slow neutrons, two or three fast neutrons emerge alongside two medium heavy fragments, making it possible to cause a fresh fission with these neutrons and trigger a chain reaction. For this purpose, the fast neutrons from fission have to be slowed down by scattering in a moderator. As early as December 6, 1939, in a report on the possibilities of energy generation from the fission of uranium, Heisenberg documented his results concerning the theory of a chain reaction. In his report, he wrote of a uranium machine with natural uranium and heavy water or "very pure carbon" as a moderator for slowing the neutrons. In this, Heisenberg anticipated the idea of a nuclear reactor.

The work in Berlin was not satisfying to him: "A lot of talk, and very little real work" took place there, he wrote to his wife in Munich in July 1940. In September, as representative of the Leipzig Academy, he had to:

> *…take part in a session of the Reich's Association of Academicians; I have never before been so horrified by the German professors as in this session. And this shamelessly arrogant tone of the Prussians into the bargain – I just wanted to run away. Had I not known that there were other people here, I would have to think that truly, time and culture had never advanced beyond the limits of the Roman limes* [the northern frontier wall of the Roman Empire in Germany].

One year later, he reported, "I idle away my time with odds and ends; there is also no subject on which a person can quite organize his energies…. It's back to Berlin tomorrow, although I'll just sit around there again, and do no meaningful work."

And in May, 1943, from Berlin:

> *Things will probably go badly for both of us till the end of the war. You, because there's no end to your work, and me, because I am alone, and have to do work that for the most part makes no sense….. If need be, I would leave all my work here without a second thought. Basically, it is utterly trivial anyway.*

Then in July: "The days go by here, in which we do any kind of useless work, in the Institute or outside it, it's always the same."

In the years from 1940 to 1942, Heisenberg tried through experiments in Leipzig with neutrons and uranium to discover how one can trigger a chain reaction in order to make a heat-generating machine, an experimental reactor. For this, the fast neutrons released in fission had to be slowed down. The slower they were, the more quickly they could be captured by a uranium nucleus and excite it to fission. In these experiments, his assistants were the married couple Robert and Klara Döpel. In July 1940, he reported that he had learned how to cement metal tubes together airtight. The purpose was to make counter tubes for neutrons. "The opportunity to learn the basics of experimental physics is quite enjoyable for me," he wrote to Elisabeth, although he did not think of himself as an experimental physicist. He was not to be compared to a master of the art of experimentation, such as Enrico Fermi.

Heavy water and highly purified graphite presented themselves as possible substances, called "moderators," for the slowing of neutrons. The suitability of graphite was investigated by Walther Bothe and Peter Jensen at the Kaiser-Wilhelm Institute for Medical Research in Heidelberg. In January 1940, they documented their results for the absorption of neutrons in graphite, which they concluded was not suitable for a reactor. They were aware in principle that the graphite they were using could contain a tiny amount of impurity, boron, which would be responsible for the problems they were having. But their measurements did not show this clearly. They resulted in a value for the absorption of neutrons in graphite which was two times higher than the real value. Because of the results, graphite was ruled out as a moderator. So heavy water remained as the only moderator.

Early in 1942, using a small spherical arrangement of layers of natural uranium and heavy water, Heisenberg was able to measure a slight increase in the neutron flux inside the reactor when neutrons were irradiated from outside. In the same year in Chicago, Enrico Fermi succeeded in triggering a self-sustaining chain reaction in a sufficiently large reactor with a moderator of super pure graphite, which the German Uranium Club of course did not know. Once Einstein's letter to Roosevelt had led to establishment of an action committee for the uranium project, Fermi received funds from the army sufficient to acquire the necessary uranium and the highly purified graphite.

Around the middle of 1941, it was clear to Heisenberg that building a bomb was possible in principle, but that the difficulties of enriching fissionable U-235, and of the production and separation of plutonium, were so great that realization of such a project in Germany was out of the question.

Nonetheless, the question remained whether in the far future a bomb could be built, specifically in the United States, with its overwhelming industrial capacity, and whether it could be dropped on Germany. Could such a development be stopped by a small community of nuclear physicists' agreeing to refuse to build such a bomb?

For nearly 20 years, Niels Bohr had been Heisenberg's great father figure and his mentor, with whom he could discuss the most sensitive matters. He therefore thought he could best discuss these questions with Bohr. He was naïve enough to assume nothing had changed in their close friendship because of the war. But that was an error: Germany had occupied Denmark. For Bohr, Heisenberg represented the enemy country, and besides, Bohr had Jewish ancestors and had to fear for his life.

Heisenberg applied to travel to German-occupied Copenhagen to attend a conference on cosmic rays organized by the German Cultural Institute in Copenhagen. Bohr boycotted this institute, but the session offered Heisenberg the opportunity of speaking with him. He met him at his institute, at home, and on a walk in the park, where no surveillance by Gestapo spies was to be feared. Under no circumstances could Heisenberg openly address his work with the uranium project. That would have constituted high treason. For that reason, he spoke in indirect allusions and hoped Bohr would understand them to mean that German physicists – and he himself – had no intention of building an atomic bomb.

However, Bohr's reaction was curt. As soon as Heisenberg began to speak of the situation, he broke off the conversation, and the mission was aborted. Bohr could not believe that Heisenberg was acting on his own initiative, and presumed he had been sent by some German government body to sound him out. The result was that on his next trip to America and Los Alamos in 1942, Bohr reported the visit and delivered the message that the Germans had a uranium program. He also showed around a sketch that was given to him by the young physicist J. H. D. Jensen from Hamburg. The drawing was given to Edward Teller, Hans A. Bethe, and Fermi. It was in the shape of a hat. Teller, Bethe and Fermi immediately recognized it as a nuclear reactor. On seeing the sketch, Fermi exclaimed, "Well they can't drop a reactor on London!"

On June 4, 1942, Albert Vogler, president of the Kaiser-Wilhelm Society, called for a session at Harnack House in Berlin to test the usability of the uranium project for military purposes. Invited were scientists of the uranium project, among them Otto Hahn, Werner Heisenberg, Kurt Diebner, Paul Harteck, Carl-Friedrich von Weizsäcker, and Karl Wirtz. Additionally, high-ranking military men and ministerial officials took part, at their head Reich

Minister for Armament and Munitions, Albert Speer. Heisenberg lectured on the utilization of nuclear fission for the generation of energy and for military purposes. The physicists were asked if a bomb could be built with uranium. For that, enrichment of the fissionable uranium isotope U-235 contained in natural uranium at just 0.7%, and the chemical separation of fissionable plutonium, would have been necessary. Both projects would require many years and a huge financial outlay, which lay beyond conceivable possibilities.

When Speer asked the physicists how much money they needed, it was none other than the theoretician von Weizsäcker who hazarded a number: 43,000 Reichsmark. Speer and Field Marshal Milch looked at each other in disbelief, and shook their heads at such naiveté. Speer had figured on a cost of 100 million Reichsmark. He reproached Vögler for having invited him to so useless a meeting.

Heisenberg was asked to present a more realistic proposal. He estimated a cost of 350,000 Reichsmark for personnel and materials for 1942; Weizsäcker wanted an additional 75,000 Reichsmark for the theory. Speer remained annoyed by the insignificance of the request "in such a decisively important matter," and raised the budgetary approach to 1 to 2 million Reichsmark.

What was the reason for Heisenberg's tendency to minimize the project? He must have seen the enormous expense in personnel, resources, and time which the extraction of fissionable material by isotope separation would require, and shrank from the huge responsibility. It was not even clear which separation process for U-235 would prove successful. He knew that Hitler dismissed all weapons projects whose realization would take more than a year. Heisenberg judged the time it would take to research the development of a bomb realistically to be at least two years up to the point of technical implementation.

After further deliberation, in response to a renewed inquiry by Speer, he increased the time to three to four years. Speer accepted this judgment, and thus in the fall of 1942 it was decided to waive the development of an atomic bomb. Instead, Speer approved the development of "an energy-generating uranium engine" for the navy's submarines. Thereupon, the Kaiser-Wilhelm Institute for Physics was fully subsumed again under the Kaiser-Wilhelm Society and dismissed from the control by the Army Ordnance Department. The physicists involved could once more presume they would not be drafted into military service at the front. Strictly speaking, this was the reason Carl-Friedrich von Weizsäcker remained on the uranium project, although he did not work on the experiments.

Two weeks after the June 4, 1942, session General Field Marshall Milch directed mass production of the V1 rocket to begin. This huge project of the

Peenemünde group around Wernher von Braun gobbled up all available means of production. Plans for the uranium reactor were put on the back burner.

Kurt Diebner left the institute and went with several colleagues to the army lab at Gottow, where he carried out several experiments in producing a uranium reactor in competition with Heisenberg. Heisenberg assumed his position in June 1942 as a replacement for the vacationing Debye; he was named director of the KWI and leader of uranium research in Berlin. The group was comprised of no more than a dozen scientists, among them Horst Korsching, Karl Wirtz, Erich Bagge, and Fritz Bopp. Carl-Friedrich von Weizsäcker accepted the professorship for theoretical physics at the "Reich" University of Straßburg in German-occupied Alsace.

In October 1943, Heisenberg had a long conversation with him, about which he reported to his wife:

> *Basically, I don't understand him at all. This way of taking everything on principle, and always forcing the 'ultimate decision' is entirely foreign to me. Weizsäcker can say things such as, for instance, that from the experience of guilt and punishment people would be ready for a new way of thinking – by which he means the new faith, to which he himself subscribes.... In discussions like that, I am fiercely opposed to him.*

This is a clear statement of how Heisenberg distanced himself politically from Weizsäcker.

However, Heisenberg did not regard the experimental and administrative work on the uranium reactor that was now delegated to him at the KWI as his main interest. In June 1942, he complained in a letter to Elisabeth that he "was [not] getting to his own work." His "own work," which was so important to him, consisted on one hand of a new theory of elementary particles based on the concept of a scattering matrix, or S-matrix, and on the other of an investigation and interpretation of the phenomena of cosmic radiation. He lectured on these topics on a trip to Switzerland in Zürich, Basel, and Bern in November 1942. Other trips on which he reported about his "own work" included a visit to Budapest in April 1941, to Pressburg in March 1943, and to Utrecht and Leyden in October 1943. In 1941 and 1942, Heisenberg led a seminar on research into cosmic radiation. In honor of Arnold Sommerfeld's 75th birthday, he assembled a festschrift of papers on this area ["Cosmic Radiation"], the first edition of which was destroyed in an air raid.

Whenever he could, Heisenberg participated in the sessions of the Wednesday Society in Berlin, a loose community of prestigious scientists, physicians, artists, military men, and high officials. They met at alternating homes of the members to hear lectures – for instance at the home of the surgeon Ferdinand Sauerbruch. On July 18, 1944, the society met at Heisenberg's apartment. Two days later, on July 20, 1944, the Stauffenberg attempt on Hitler's life in the Wolf's Lair took place. Numerous members of the Wednesday Society were arrested as supporters of the assassination attempt and executed.

The Uranium Reactor

While in 1942 Enrico Fermi had already observed a critical chain reaction in Chicago, further progress of the German project proved extraordinarily difficult. There was not enough uranium in Germany, and furthermore, three different groups were competing for what little there was – apart from the Berlin KWI, the Hamburg group under Paul Harteck and Kurt Diebner's Gottow group. Altogether, this was a few dozen scientists and technicians. Because of the results of absorption measurements of contaminated graphite, only heavy water was used to slow the neutrons. But acquisition of this substance was difficult, it was as expensive as gold, and it was the lack of heavy water, ultimately, that would prove fatal.

The only producer in the world was the Norwegian firm Norsk Hydro, in Vemork, which used the abundant available hydropower to achieve accumulation of heavy water by distillation. Norway had been occupied by German troops since June 1940. In February 1943, a Norwegian-British commando raid smashed the cells inside the guarded factory, destroying the stocks of heavy water. After the reconstruction of the plant, in November 1943, a squadron of 140 American B-19 bombers wrecked the facility totally. The remaining heavy water was to be shipped to Germany, but in February 1944, a Norwegian resistance group, with British help, succeeded in planting explosives on the ferry which was to transport it. The boat sank with all crew and passengers. In Kurt Diebner's view, the destruction of the heavy water plant and of the supply of heavy water was the principal reason for failure of the reactor project. In the summer of 1944, because of the air raids on Berlin, the KWI for physics and the KWI for chemistry were relocated to Hechingen and Tailfingen in Württemberg, respectively. The uranium reactor project followed at the end of 1944, and was to be reassembled in a former wine cellar beneath the castle Church of Haigerloch, close to Hechingen.

At the beginning of 1945, the reactor at Haigerloch did not quite reach the critical state, i.e., the condition in which a self-maintaining reaction would be triggered. In any case, though, the amount of heavy water on hand was insufficient. Similarly, Diebner's group in Gottow achieved a doubling of the neutron flux in their setup, but no self-sustaining chain reaction. There, too, heavy water was in insufficient supply.

In the meantime, Heisenberg took time off to pursue "his" theoretical physics work, as well as to play chamber music. At the end of 1944, he wrote, "I've even made significant progress. It's actually absurd under these circumstances still to want to do science. But nevertheless, I find it beautiful. I make music with pedantic exactingness as well."

In the spring of 1945, he even performed in a public concert as pianist in Hechingen.

It is clear from a letter Heisenberg wrote to his wife on February 1, 1945, that from the viewpoint of the scientists involved, the project also served the purpose of saving them and their younger collaborators from military service. "The inner struggle in our Nuclear Physicists' Society [Diebner in Thüringen against the KWI] has been reignited, which is no doubt related to the threat of the draft, and the looming peril from the East."

As the western front drew ever nearer, Heisenberg arranged to have the blocks of uranium buried, and in April 1945, he set out on his bicycle on the 270-km journey to join his family in the Bavarian village of Urfeld.

Late in April 1945, members of an American special unit of the Alsos mission, under the command of Colonel Boris T. Pash and the physicist Samuel A. Goudsmit, reached Hechingen and Haigerloch shortly before the advancing French troops. Goudsmit was astonished at how small and unprepossessing the whole experimental setups were. In his book *Alsos* he wrote:

> *It was so obvious that the whole German uranium setup was on a ludicrously small scale. Here was the central group of laboratories, and all it amounted to was a little underground cave, a wing of a small textile factory, a few rooms in an old brewery. To be sure, the laboratories were well equipped, but compared to what we were doing in the United States it was still small-time stuff. Sometimes we wondered if our government had not spent more money on our intelligence mission than the Germans had spent on their whole project.*

At its high point, the American Manhattan Project employed approximately 150,000 people, the German reactor project, a few dozen.

The German scientists still present at Hechingen – Bagge, Korsching, von Laue, von Weizsäcker, and Wirtz – were taken prisoner by the Americans. They were also looking for Heisenberg but learned that he had fled. Colonel Boris T. Pash set out with a small unit for Urfeld, where on May 3, 1945, he also arrested Heisenberg. Together with the Haigerloch scientists and Gerlach,

Diebner, Harteck, and Otto Hahn, he was taken via a few detours to Farm Hall in England and interned there. The conversations among the ten physicists were taped by the British secret service. When, on August 6, 1945, the news of the devastating bombing of Hiroshima spread, the German physicists at first didn´t believe it was an atomic bomb. Then they began to believe it.

Hans A. Bethe, the director of the theory department of the American bomb project at Los Alamos, concluded from the protocols that Heisenberg did not know how to calculate the critical mass at this time, indicating that he did not work on atomic bombs during the war. Edward Teller, the father of the hydrogen bomb, judged that Heisenberg had sabotaged the bomb; he saw conclusive evidence that Heisenberg did not put any real effort into the development of it.

The Americans dismantled the Haigerloch reactor and brought the entire apparatus with the uranium they found, and the heavy water and countless files, including the private scientific letters to Heisenberg from Madame Curie, Bohr, and Einstein to the United States. Heisenberg summarized the work of the Uranium Club thus:

> *During the war, we were lucky that working on the nuclear arming in the war proved impossible, because it would have taken too long. I could quite honestly report: in principle, it is possible to build atomic bombs, but all the processes that we know up to now are so terribly expensive that it would take years and require a quite enormous technical expenditure of billions.*

6

Social Affinities

Einstein's Women

Einstein's attractive looks and his stimulating and witty conversation appealed greatly to women. Already in his high school years in Aarau, Marie Winteler, the 16-year old daughter of his teacher and landlord, fell in love with him. She was two years older and thought of herself as his "little, insignificant darling," who could not compare herself to Albert intellectually. He, on the other hand, gladly exploited her affection; for example, soon after his departure from Aarau to Zürich, he would send her his dirty laundry for her to wash and send back. But the more he settled in at Zürich, the more he gave precedence to study and science, and he broke off the relationship and avoided further visits so as not to lead Marie on.

Another reason he broke it off was his friendship with the student Mileva Marić, three and a half years his senior from Novi Sad, in Serbia, at the time part of the Austro-Hungarian empire. He could discuss subjects in physics with her. He valued her as his "equal, strong and independent," and soon his passion for her burst into flame. Their correspondence from this period, from 1897 to 1903, is filled with tender declarations of love. One of Mileva's character traits came to the fore in these years – she was highly jealous of any other woman Albert had a relationship with.

When Mileva spent a semester studying in Heidelberg, he wrote the "honorable Miss" a four-page letter in which he asked her to write him any time she was bored. She replied almost poetically that just then she was strolling under German oaks in the lovely Neckar Valley, but whose "charms were shamefully enveloped in dense fog." Her father in Serbia had given her tobacco

K. Kleinknecht, *Einstein and Heisenberg*, Springer Biographies,
https://doi.org/10.1007/978-3-030-05264-5_6

to bring along to Albert; he "wanted so much to make your mouth water for our little robber country…. You really must come along sometime." At the end of the letter, she refers to a "nice" lecture by Philipp Lenard on the velocity and mean free path of molecules.

Albert's reply and all subsequent letters resembled Mileva's – a mix of amusing and stimulating observations and stories about lectures and seminars. Then he offered the advice "to come here as soon as possible." She would not regret it. And whatever substance she missed from the lectures, she could make up with "our" transcripts, by which he meant his own and those of his friend Marcel Grossmann. The letter succeeded. Mileva returned after one semester. From then on, they were often together. In one letter, for instance, he announced his coming simply: "If agreeable to you, I'll come to read this evening at your place."

At home in Milan with his parents, he showed his mother a photo of Mileva; she was impressed at first. "Oh, but she is a clever minx." He sent her greetings "from my wrinkly." But his mother's initial impression of Mileva would soon change, because she did not live up to Pauline's idea of an ideal daughter-in-law. Albert in turn thought his mother a bit narrow-minded and "Philistine," his favorite expression for provincialism, which he had adopted from Schopenhauer. He thought it curious that, "With time, the closest natural bonds in the family gradually degenerate into friendships of habit, and we are mutually so inwardly unknowable that it is impossible for one person to feel empathetically what moves the other."

But he understood his dear little Doxerl splendidly. In August 1899, he wrote from his summer vacation in his family circle in Mettmennstetten near Zürich that in the happy tranquility of vacation, study was a diversion, "not the lazing about we're accustomed to in our household." Apparently, they were already occasionally living together. The letter goes on, "You are a really clever girl (Hauptkerl)[1], and pack a lot of vitality and health in your little body." In Swabian dialect, *Hauptkerl* is meant as the highest praise. The obligatory excursus into physics follows, with fundamental insights prefiguring the year 1905:

I am more and more convinced that the electrodynamics of bodies in motion, as currently presented, does not correspond to the truth, but rather could be presented more simply. The introduction of the name 'ether' into the electrical theories has led to the idea of a medium, of whose motion one can speak without – as I believe – being able

[1] An arcane, now rarely used word: a person (implicitly male) who is or is seen as very capable or smart.

to connect this statement with a physical meaning. I believe that electrical forces are directly definable only for empty space – also stressed by [Heinrich] *Hertz.*

We see here that Maxwell's electrodynamics, which he had learned as a high school student in Munich from his Uncle Jakob, was a constant companion, and that he was prepared to discard the idea of an ether once and for all. In a paper by Wilhelm Wien, he read about the experiment of Michelson and Morley in which the ether hypothesis was refuted. Next, he complained about the mindless chatter of his friends and relatives who came to visit, especially an aunt from Genoa, "a veritable monster of arrogance and mindless formalism." But he and Mileva, by contrast, "understand each other so well with our black souls, and also in our coffee drinking and sausage eating." After the vacation, Einstein moved on his return to Zürich, near her but not into her house, "for the sake of people's wagging tongues."

When Albert accompanied his sister Maja to Aarau at the beginning of her studies at the teachers' seminar, he was careful not to fall in love again with Marie Winteler, for "were I to see that girl a few times, I would certainly go crazy – that I know, and fear it like fire."

After he successfully completed his teaching exam in July 1900, Albert traveled to his mother and sister in their Swiss summer resort in the Melchtal. He had decided to marry his Doxerl, and discussed his plans beforehand with Maja. On his arrival, his mother asked quite benignly, "Well, and what will become of Doxerl?" Her son's laconic reply, "My wife," elicited a dramatic scene. His mother wept, complaining that he was wrecking his future and cutting off his life's journey, because Mileva could not marry "into any respectable family." She was bookish like him, and when he was 30, she would be "an old witch." The summer resort became a torture for him, since he had to "flatter" acquaintances and relatives of the family.

The Mileva dispute swelled further, but was covered over by Albert's entertaining the vacationing guests musically. Albert was also confident that he would prevail, because his parents were apathetic and had in their entire being less stubbornness than he had in his little finger. And his love was unaffected by his parents' resistance. He wrote his Doxerl from the resort that if he could not have her he would feel incomplete.

Following his return to Zürich, Einstein looked for a position. He submitted applications for an assistantship at technical institutes and universities in Switzerland, in Germany, and in Italy. But neither with his diploma advisor Professor Weber, nor with the mathematical physicist Emanuel Hurwitz, did he have any luck. He wrote to Wilhelm Ostwald, to Eduard Riecke in Göttingen, to the Polytechnic in Stuttgart, and tried through his father's friend Michele Besso to get a position in Milan, all without success. "I will soon have honored all the physicists from the North Sea to the southern tip of Italy with my offer," he wrote Mileva. Not until April 1901 did the fog lift with the offer to accept a two-month stand-in position at the Polytechnic in Winterthur. During this time Mileva, who had failed her teaching exam, prepared in Zürich to take the exam for the second time in July 1901 – once more, unsuccessfully.

In April, she had confirmed that she was pregnant, although in their correspondence, Albert asked only near the end of May for the first time after the "boy," the "little son, and your doctoral paper." From time to time he wrote enthusiastically about a paper by Philipp Lenard from Heidelberg on the emission of cathode rays (electrons) from metal electrodes by ultraviolet light. Reading this paper "filled him with such happiness and such pleasure" that he absolutely wanted to let Mileva share in it.

In 1901, after failing the exam for the second time, she went home to Novi Sad. There, she had a shock. Albert's parents had written Mileva's parents a letter in which they flatly rejected the marriage. Now she was worried. Would Albert keep his promise nonetheless? In the meantime, he had accepted a position at a private school in Schaffhausen, at the Rhine Falls, to prepare a pupil for the graduation certificate. So Mileva struck out to visit him there and reassure herself of his affections. Lest she cause a stir, though, she spent the nights in the neighboring Stein on the Rhine, where she stayed for more than two weeks. She visited him in Schaffhausen and received his return visits at Stein on the Rhine. If he failed to come, she was "momentarily quite angry."

After two weeks in Stein on the Rhine, she returned to Novi Sad to prepare for the birth of her child at home with her parents. She was hoping for a girl, whom she called Lieserl; he rather imagined a boy, quite in secret ("so that Doxerl doesn't notice") named Hanserl.

The only problem to be solved, he wrote in November, is "the question how we can bring our Lieserl to us; I would not like to have to give her up."

In January 1902, the child was born in Novi Sad. A year and a half later it was mentioned in a letter for the last time, since it had fallen ill of scarlet fever. No further documents are to be found. To the end of his life, Einstein hid the existence of his daughter and swore his executors to secrecy. Only after Mileva's death in 1948 were the love letters from this time located by his daughter-in-law Frieda Einstein-Knecht at the closing of Mileva's apartment on Huttenstraße in Zürich. Because of the opposition of Einstein's executors, they could not be made public until 1987 through the action of Robert Schulmann and Jürgen Renn. Most historians assume that Lieserl was given up for adoption in Novi Sad. In any case, proof of this is lacking.

However, a fresh trail popped up in 2015 from letters to the present author from Helmut Lang, a German now living in Canada. From his birth in 1956 until 1974, Lang lived with his grandmother, Marta Zolg, née Gießler, in Bietingen, a district of Gottmadingen near Constance. He learned from her that she was Einstein's daughter. According to her story, the child born as Marta Marić in January 1902 in Novi Sad was taken by carriage in December 1903 to Nordrach near Oberhamersbach in the Black Forest, and there registered on January 7, 1904, as the daughter, Marta, of the farmer Michael Gießler (1864-1928) and his wife. After an apprenticeship at a hotel named "Prestige" in Frankfurt from 1918 until 1928, where she learned that Albert (she called him the "higher animal") was her father, and that her parents had given her up for adoption, she lived in Bietingen with her husband, Ernst Zolg, and had three children. In 1980, she died at the age of 78 in Bietingen.

Mileva and Albert married in 1903 without the participation of either family. His father, Hermann Einstein, gave his consent just before his death on October 10, 1902.

But for the young wife, the trauma of having to give up her child came as a shock, which had aftereffects and strained the young marriage, as her son Hans Albert (born in 1904) later indicated. Another son, Eduard, born in 1910, suffered health and mental problems that caused his mother great stress.

Then, Einstein's meteoric rise to becoming a world-renowned physicist began. Mileva cared for household and children. She accompanied him along the various stations of his rise uncomplainingly: Bern, Zürich, Prague, Zürich once more. Here, she felt at home. When Einstein accepted the attractive offer from Berlin, the couple's alienation from each other increased. Mileva did try out a move to Berlin but felt isolated there and could not really imagine living there. She had noticed that for some time Albert had had a relationship with his cousin Elsa in Berlin. To Elsa he had conceded he would give much to be able to spend a few days with her, but without his "cross," by which he meant Mileva, whom he later abbreviated with a "†" sign, and called the "sourest sourpuss." Albert thought at first of a *ménage à trois* with Mileva as wife and Elsa as lover. Already in November 1913 however, Elsa wanted him to get a divorce.

Fig. 6.1 Einstein with Elsa and her daughter Margot, 1927

In March 1914, when Einstein took up his position at the Academy, he had to travel alone. Mileva balked at the move and succumbed to depression. She knew no one in Berlin, but she knew about Elsa, and the intended move of her critical mother-in-law Pauline to Berlin also frightened her. So, on the advice of her doctors she went first to Locarno to recover before she followed to Berlin with the children in May. She did not want to get acquainted with Albert's new colleagues, Planck, Nernst, and Haber.

Einstein had no patience for her shyness and insecurity. Nothing came of the plans for a family life together in an apartment in Dahlem. Both finally moved out. At this point, Albert had inwardly already separated from Mileva, and he made this clear to her in July by setting out written rules of behavior. He presented to her the conditions under which he would be prepared to live together with her again. In addition to maintaining his clothes and his laundry, as well as his study and bedroom, he required that three meals daily be served him in his room. Mileva was to give up all personal relations to him, except so far as keeping up appearances was required in social situations. Neither tenders of affection nor reproaches were permitted, and she must leave his room immediately if he asked it.

The separation in the summer of 1914 was the result of their muddled relationship. Accompanied by her friend Michele Besso, who had arrived from Zürich, Mileva departed Berlin with the children. Weeping, Einstein saw them off at the station. But Albert thought he had found happiness with his cousin Elsa, with whom he had been intimate since childhood. The pain of saying farewell was quickly overcome. After his leavetaking from his family at the station, he wrote her:

> *You* [dear] *little Elsa will now become my wife, and be convinced that it is not at all hard living with me. I know you understand that. After so many years, you will once again run the household and do as you like, and everyone will show you respect.... Write soon, and be warmly kissed by your Albert. Greetings to Ilse & Margot from their step-father!*

To begin with, though, he felt indescribable pleasure in knowing he had regained his "Castle Peace-of-Mind" and was living tranquilly and alone in his large apartment. "The decision to isolate myself, comes as a blessing," he wrote his friend Ehrenfest.

Aside from the weekly Wednesday sessions at the physics colloquium, Thursdays at the Academy, and fortnightly meetings at the Physical Society, he now worked undisturbed in solitude on the general relativity theory, and fell into a frenzy of creation, in which he ignored his bodily needs, ate irregularly, and worked through the nights.

The outbreak of war and popular enthusiasm for it left him a pacifistic outsider. "It is unbelievable what Europe has now begun in its madness," he wrote to Ehrenfest. But he cheered himself with sarcasm: "Why shouldn't one be able to live contentedly as a servant in the madhouse?"

And he threw himself into his work, even though colleagues thought the problem of gravitation was unsolvable. He had now been working on it for seven years, and "my sequence of papers on gravitation is a series of false paths," as he later conceded. In October 1915 he determined that the latest of his sessions-reports to the Prussian Academy publishing his results were untenable, and that he was compelled to return to his field equations developed earlier. These now proved to be the correct ones, as described earlier.

It was a superb success. He wrote that he "had worked like a horse, smoked like a chimney, and eaten hardly anything." Food became scarcer, too. In December he confessed to his friend Besso that he was "satisfied, but fairly broken."

The creative frenzy continued. In 1916, two publications followed, on gravitational waves, the Schwarzschild solution to the cosmological equation, and the Einstein/de Haas effect. But success took its toll. Einstein fell ill, had digestive and liver problems, and had to shift to a new diet. Not even food packages sent by relatives in Switzerland helped. So Elsa, hitherto able to help only from a distance, now intervened.

In the summer of 1917, she persuaded him to move from his previous apartment to one near her. She lived with her two daughters in her parents' house at 5 Haberlandstraße on the ground floor. From there, after his move, she could cook for him, when from December on he was confined to his bed for months because of a stomach ulcer. Not till April 1918 could he go out again. By May 1918, he had not yet recovered from the jaundice he had contracted. Now Elsa urged him ever more determinedly to finally divorce Mileva.

After repeated attempts, which were regularly rejected by Mileva, the third one finally hit its mark. In the summer of 1918, after lengthy negotiations over financial arrangements, Mileva agreed to the divorce. In the agreement, she was awarded custody of the children and the money for the future Nobel Prize, on which Einstein certainly counted. A hearing in a Zürich court followed in February 1919, at which Mileva sued Albert for divorce on the grounds of adultery. The court approved the agreement and imposed on Albert, who had been declared guilty, the condition that he was not to marry for two years. Mileva remained in Zürich with the sons, Hans Albert and Eduard. In 1922, when Einstein received the Nobel Prize, he transferred to her (presumably "to save the taxes") only one quarter of the money, which she used to purchase the house at 62 Huttenstraße. Seven years later, he transferred another quarter. He invested the remainder with an American bank, Mileva receiving the interest. Management of the account lay in his hands. Mileva lived at the Huttenstraße address until her death in 1948.

Four months after the divorce proceedings in Zürich, Einstein married Elsa in Berlin. She was 43 at the time, he, 40. The couple moved with Elsa's daughters, Ilse and Margot, into Elsa's apartment at 5 Haberlandstraße. Two more rooms were set up as studies for Einstein, one of them the "tower room." In many ways, Elsa represented Albert's highly disdained bourgeoisie, the Philistines, as he called them, following Schopenhauer. She mothered her "little Albert," and was proud of him and of being the wife of an important man.

Charlie Chaplin observed that she was a square-built woman, who was happy to be the wife of a great man and made no secret of it. Under her care, his health improved, as he wrote Besso in 1920. He let himself be spoiled and happily received guests in the apartment, a bohemian in a bourgeois household.

Fig. 6.2 Einstein in the tower room of 5 Haberlandstraße, Berlin, 1927]

For Elsa, life with Albert had its drawbacks, to be sure. Work was his priority; family life was unimportant. Elsa wrote, "He goes into his study, comes down, plays a few chords on the piano, wolfs something down, goes back to his study. On days like that, Margot and I make ourselves scarce. We put something to eat on the table, and lay out his coat (in case he should like to go out)."

Disillusioned, she wrote to Ehrenfest that Albert's obsession with work was bottomless, incomprehensible. A guest in the house remarked that one did not get the impression there was much intimate affection between the two. Elsa's bedroom was next to that of her daughters; Albert's bedroom, on the other hand, was at the end of the hall. Not much changed in Albert's disdainful attitude towards the marriage during its course. His comment about it sounds cynical: "Living is like smoking, specifically marriage." Einstein knew that he was not cut out for marriage. On the occasion of a eulogy of his friend Michele Besso, he wrote, "What I admire in him as a human being is that he managed to live many years in peace and harmony with a woman – an enterprise in which I have twice failed scandalously." Later, he expressed it even more negatively: I have survived two wives and the Nazis.

His sense of freedom increased when he moved into his summer house in Caputh on Schwielowsee west of Berlin and spent the summer months sailing and swimming. He entertained female friends there, Estella Katzenellenbogen, Betty Neumann, Grete Lebach, Johanna Fantova. Elsa stayed in the city.

Elsa died in 1936, three years after they emigrated to the United States and year after they had moved into a house in Princeton. Another daughter was born to Albert in 1942, stemming from Einstein's relationship with a nightclub dancer in New York, as his stepdaughter Margot wrote to Hedwig Born after his death in 1955. This daughter was also given up for adoption. Science held precedence.

Heisenberg's Family

Werner Heisenberg enjoyed a sheltered childhood together with his brother, Erwin, in Würzburg and Munich. Throughout his school and university years, he focused on the work – girls played no role for him. His romantic inclinations played out in his youth group during hiking tours in the mountains or in the Altmühl Valley, always associated with music and philosophical discussions. Close friendships tied him to students his age and hiking companions. Experiencing nature in the open, and understanding its inner connections was his goal. That is how he wished to live.

After the Abitur, he began his studies with Sommerfeld in Munich. Unusually young, still only 20, he completed his doctoral studies and stood for the oral defense. There was no time, and perhaps no interest, in finding love. Whereas his friend Wolfgang Pauli spent his evenings in bars, Heisenberg was an early riser. The doctorate in Munich was followed by his assistantship with Max Born in Göttingen; his reports to his parents offer information about it. His first meeting with Niels Bohr, his highly esteemed role model at the "Bohr Festspiele" in Göttingen; his sojourn in Copenhagen; his habilitation in 1924 in Göttingen; his breakthrough to quantum mechanics on Helgoland in 1925; his insight into the uncertainty principle in Copenhagen in 1927 – all this was concentrated into just a few years. A professorship in Leipzig in 1928 was the logical outcome of this great success, and it was naturally crowned with the awarding of the Nobel Prize in October 1933 for the year 1932. His mother accompanied him to the Nobel ceremony in Stockholm. He was 31, and still unmarried.

It was only now that he seemingly began to be interested in women. His doctoral student Carl-Friedrich von Weizsäcker, ten years his junior, had a younger sister, Adelheid, with whom Werner fell in love. But her parents apparently regarded the middle-class Heisenberg as not on a par with her, and besides, Adelheid was just 17. It was only several years later that music paved the way to his wife. In January 1938, during an evening of chamber music in Leipzig, he played the piano part of Beethoven's 2nd trio. During the slow movement, *Largo con espressione*, his glance met that of the aspiring bookseller Elisabeth Schumacher. In his autobiography, he writes shyly, "and the slow movement of the trio (Beethoven, G major) became from my side an extension of a conversation with this listener."

No more than two weeks later, the couple celebrated their engagement, and three months later, their wedding took place. Nine months later, the twins Anna Maria and Wolfgang were born, and in subsequent years five more children followed: Jochen (1939), Martin (1940), Barbara (1942), Christine (1944), and Verena (1950).

The couple was separated by the outbreak of war in 1939, and in the next eight years they were able to spend at most half of those days together. While he waited in Leipzig to be conscripted, was ordered to Berlin, was sent back to Leipzig, and worked on the uranium project, Elisabeth moved to the farmhouse in Urfeld atop the Walchensee with the children, and muddled through despite all the awful circumstances of wartime.

Their correspondence during the time of their separation documents strikingly what life was like under the dictatorship. Aside from the usual daily problems, they come back always to their enjoyment of literature or music, Storm, Stendhal, Schubert, and of course to the children's education. Heisenberg does not write about his research at the secret uranium project, but on the other hand he writes much about his "own science," for which he always has too little time. By this he means the problems of theoretical physics preoccupying him at this time, the scattering of elementary particles, and his research into cosmic radiation.

Their letters also offer insight into the question of why Heisenberg declined the offers from America. Even though it was foreseeable that during the looming war, life in the United States would be easier, and even though he could expect better scientific circumstances there, he could not imagine giving up his homeland and having his children grow up in a culturally different environment. Whether these were the decisive reasons as to why the couple was against emigration is not known. In any case, ultimately, together they determined to stay.

Einstein's Religion

Albert Einstein, in whose family religion played no great part, received Catholic religious instruction in grammar school and "Israelite" instruction in high school. He discovered that the biblical stories of creation were incompatible with the view of nature he had gotten from popular scientific publications, and he began to doubt the Bible. He believed only that which he understood. Moreover, he rejected any demands from an authority, including that of the religious communities. He resisted taking part in religious services and belief in any particular dogmas. Therefore, he decided already in high school in Munich to leave the Jewish religious community, which he did after his flight to Italy.

The political developments, the anti-Semitism, and the persecution of Jews eventually led to his becoming an ardent and eloquent supporter of Zionism. After the founding of the State of Israel, he was even asked to become president of the country.

Viewed superficially, one might take him to have been a religious person, as did Friedrich Dürrenmatt in a lecture in 1979 at the ETH Zürich, reflecting on his frequent mention of the name of God. "Einstein used to talk about God so much that I almost suspect he was a secret theologian." His religious convictions were independent; he was skeptical of the idea of a personal God.

In 1929, he replied to the question "Do you believe in God?" with a telegram to Herbert S. Goldstein (New York): "I believe in Spinoza's God, who reveals himself in the lawful harmony of all that exists, but not in a God who concerns himself with the fates and the doings of mankind."

This conception of a God who reveals himself in natural laws corresponds to the old idea of a "measuring" God, who measures Earth's radius with compasses, as depicted by a medieval book illustrator in Reims in 1250. In this view, Einstein and Heisenberg approach each other pretty nearly.

Fig. 6.3 The measuring God, Bible Moralisée, 1250, Reims

When this telegram became known, strong opposition to him arose in America. Particularly his rejection of a personal God elicited the protest of pious Americans. But he did not retreat from his opinion and argued theologically:

> *If this personal being is omnipotent, then every event everywhere in the universe is his work – including every human action, every human thought, every human feeling. So how is it possible to think of holding people responsible for their deeds and thoughts before such an almighty being?.... In giving out punishments and rewards, he is in a way passing judgment on himself.... How can this be combined with the goodness and righteousness ascribed to him?*

Much later, a year before his death, Einstein addressed the urgent questions of the writer and philosopher Eric Gutkind, who had sent him his 1952 book, *Choose Life: The Biblical Call to Revolt.* This Zionist-oriented book interprets Judaism as the avant-garde of humankind. Einstein's letter to Gutkind is available on the Internet. In 2008, it was sold by Bloomsbury Auctions, and in

2013 sold again on eBay for more than $3 million. In this letter (translated from the German original to English), he lays out comprehensively what he thinks of Gutkind's idea of God in particular, and of religion in general:

> *Princeton, 3.I.54. Dear Mr. Gutkind! Encouraged by* [Luitzen Egbertus Jan] *Brouwer's repeated suggestion, I have read a great deal in your book in the last few days, for whose sending I thank you very much. What struck me especially was this: regarding the factual outlook on life and on the human community we are largely in agreement: about a personal ideal of striving towards liberation from egocentric desires, striving towards improvement and ennoblement of existence, with stress on the purely human, whereby the lifeless thing is to be regarded only as a means, that must not be conceded any dominant function. (This position as holding a genuinely 'un-American attitude' is especially what connects us.) Nevertheless, without Brouwer's encouragement I would never have brought myself to delve into your book so deeply, because it is written in language that is inaccessible to me. The word God is for me nothing more than the expression and product of human weaknesses, the Bible a collection of honorable, but still primitive legends. No interpretation however subtle can (for me) change any of this. These refined interpretations are by their nature highly diverse, and have almost nothing to do with the original text. For me the pure Jewish religion like all other religions is an incarnation of the most childish superstition. And the Jewish people to whom I gladly belong and with whose ethos I have a deep affinity have no different dignity for me than all other people. As far as my experience goes, they are also no better than other human groups, although they are protected from the worst excesses by a lack of power. Otherwise I cannot see anything 'chosen' about them. Actually, I am pained that you claim a privileged position and seek to defend it by two walls of pride, an external one as a human being and an internal one as a Jew. As a human being you claim, as it were, an exemption from causality otherwise accepted, as a Jew the prerogative of monotheism. But a limited causality is no longer causality at all, as our wonderful Spinoza first recognized so insightfully. And the animistic conceptions of the nature religions are in principle not revoked by monopolization. With such walls we can only attain to a certain self-deception, but our moral efforts are not advanced by them. Rather just the reverse.*
>
> *Now that I have quite openly stated the intellectual differences in our convictions, it is still clear to me that we are quite close to each other in essential things, i.e., in our evaluations of human behavior. What separates us are only intellectual accessories, or 'rationalization' in Freudian language. Therefore I think we would understand each other quite well if we discuss concrete things.*
>
> *With friendly thanks and best wishes,*
> *Yours, A. Einstein"*

Two months later, in March 1954, he repeated this credo in other words: "In the event there is something in me that could be called religious, it is an unlimited marveling at the structure of the world, as far as our science can reveal it."

Heisenberg's Religious Philosophy

Heisenberg grew up in a Protestant family. He attended Evangelical religious instruction, was confirmed, and was married in the Church. He has written about his religious beliefs in his book, *The Part and the Whole*. At a discussion of the positivist philosophers with Bohr and Pauli in Copenhagen in 1952, he stated:

> *We know that in matters of religion, we are dealing with a language of images and metaphors, which cannot present what is meant exactly. But ultimately, in most of the ancient religions, we are dealing with the same substance, the same facts, that the images and metaphors mean to present, and that are centrally connected to the question of values. But the task remains to understand this purpose, since it obviously stands for a decisive part of our reality.*

He goes on, "Is it completely pointless to imagine a 'consciousness' behind the organizing structures of the world as a whole, whose 'intention' this is?" Here, Heisenberg's thinking meets Einstein's, who often spoke of God as the "old man," and believes he can recognize the plan of the old man behind the wonderful and mathematically expressible natural laws. Einstein expresses his belief that these laws are not chaotic, but are transparent to us in his characteristically ironic way; thus: "God is subtle, but he is not malicious."

Heisenberg continues:

> *The question of values – that is after all the question of how we act, what we aspire to, how we should behave.... it is the question about the compass by which we should orient ourselves as we make our way through life. This compass has been given quite various names in the various religions and world-views: fortune, the will of God, meaning.... I have the impression that all the formulations are dealing with the relations of human beings to the central order of the world.*

To Pauli's question whether he believes in a personal God, Heisenberg says that for him, the central order (his symbolic word for God) can be present with the same intensity as the soul of another person. And our Western ethics still has its basis in Christianity.

In the dark year of 1942, Heisenberg wrote a comprehensive essay on the "Order of Reality," in which he classifies all the branches of the sciences and arts in a grand picture of his world view. He did not publish the paper at the time, because it contained parts that could be interpreted as a critique of the current political circumstances. It became known only after his death.

In this essay, music, too, plays an important role. His daughter Barbara Blum writes:

> *For him, music meant, like mathematics, a gate to knowledge of what he called the central order. In it, he saw the activity of the "one, to which, in the language of religion, we come into relationship with," and which, without further doubt, he experienced as the good, in contrast to everything confused and chaotic.*

All his life, he felt his mission was "to advance towards an understanding of the reality that comprehends the various interrelations as parts of a single, meaningfully ordered world."

In doing so, he was ever aware that language is ultimately inadequate: "The capacity of human beings to understand is unlimited. [But] One cannot speak of last things. In their place, music itself must speak." A friend said of Heisenberg, "Where religion begins for others, for Heisenberg it is music."

More on the Role of Music

Einstein grew up in a household in which his mother, Pauline, loved music; she herself played the piano. It will also have been his mother who sent him for violin lessons at the age of six. The basics of the violin demand first of all a hard schooling in technical studies. The young Albert didn't like these at all. And apparently, his teachers did not have the ability to bridge over the drought of the early years of practicing. Not until he was 13, when he once heard a violin sonata by Mozart, was the wish instilled in Einstein to play a sonata like that. When he realized that his technical abilities were insufficient for that, he began to practice in earnest. Later, when the famous violin virtuoso Joseph Joachim gave a concert in Aarau, Albert got hold of the music of the pieces to be played and wanted to become acquainted with them before the concert by playing them himself. So he tried out the G major Sonata of Johannes Brahms, the "Rain Sonata," op. 78, a difficult piece.

At his high school certification exam in music, he played an adagio from one of the Beethoven sonatas, which was graded by the teacher as "insightful." This shows that Einstein had acquired an appreciable ability. Curiously, it was also customary at that time to play the vocal parts of songs by, say, Franz Schubert or Robert Schumann, on the violin. Einstein appears to have appreciated particularly those Schumann songs whose lightly ironic texts had been written by Heinrich Heine.

Later, in the 1920's, in Berlin he was invited to the home of Max Planck in the Grunewald suburb. Planck, a nearly professional pianist, hosted regular chamber music sessions – sometimes with virtuosos like the violinist Joseph Joachim and sometimes with amateurs like Einstein as violinist and Planck's son Erwin as cellist –for piano trios. When famous soloists came to Berlin on their concert tours, they might wish to meet the famous Einstein. Sometimes, Einstein used the occasion to invite the musical celebrities to play chamber music with him. One of these celebrities was the Russian cellist Gregor Piatigorsky, who became principal cellist of the Berlin Philharmonic Orchestra from 1924 to 1929 under Wilhelm Furtwängler. With him and a pianist, he played a piano trio. On one occasion, when the piece was over, Einstein asked Piatigorsky how his playing had been. Piatigorsky replied, "Relatively good!"

Heisenberg's family boasted a long musical tradition. A portrait of his ancestor August Zeising, a student of Louis Spohr and a recognized violinist in his time, hung in his study. Heisenberg's father loved music; he sang the most difficult arias with enthusiasm. The young Werner began piano instruction at the age of five, and soon advanced so far that he could accompany his father in songs and operatic arias.

That he was a talented musician is clear from the repertoire that he acquired relatively quickly in his early years. While he was working on the farm in Miesbach in as part of a volunteer service in 1918, he still had enough energy to practice Liszt rhapsodies and to work up the piano part of a Grieg violin sonata.

On his 21st birthday, he wrote to his brother from Göttingen that he had played Chopin preludes all evening long and had played together with Max Born for the first time. "We played a Mozart and a Beethoven piano concerto on two pianos, i.e., so that one piano took the orchestral part. The Beethoven Concerto, which I had not known, was unbelievably beautiful." From the postscript, we can infer that he was able to read this concerto at sight.

In the years from 1932 to 1936, under the guidance of a teacher in Leipzig, he worked up several piano concertos by Beethoven, which he learned by heart. He studied counterpoint and fugues, and in 1932, even tried composing a fugue himself.

Heisenberg was especially interested in analytical work on the music, since he recognized mathematical structures similar to those in natural science. He describes it thus: "But also the representations of reality that are quite remote from the exact natural sciences, such as music or visual art, on closer analysis reveal an inner order that is very closely related to the laws of mathematics. This order can appear so clearly, as in a Bach fugue, for instance, or a symmetrically repeating ornamentation, or they can make themselves known first by a particular balance, the lucid, unmediated beauty of a melody.... [A] closer investigation always reveals simple mathematical symmetries like those dealt with by mathematicians in group theory."

In his essay "Order of Reality" he speaks of becoming conscious of the other, higher world. "This applies also – especially now in our time – for many people who belong to no religious community, and who in the notes of a Bach fugue or in the flash of a scientific intuition encounter the other world perhaps for the first time."

Heisenberg was no virtuoso; his art was his soft, sensitive touch, which came into play especially on his Blüthner grand piano, and was ideal for the interplay of chamber music.

For him, music was communication. He wrote to his parents from Copenhagen, where he found it hard to settle in, "Tonight, I will play Beethoven cello and piano sonatas with a young physicist. This will surely be fine. One really can't live without music. But when one listens to music, one sometimes hits on the absurd idea that life has a meaning."

Wherever he went, he made music with friends and colleagues in Leipzig and Berlin; with Max Born in Göttingen; at the Bohrs in Copenhagen; on one of his trips to America in 1929, with colleagues in Boston and Montreal; with Karl Klingler in Berlin, and with Denes Zsigmondy in Munich. In his family, each of the children learned to play a musical instrument – violin, viola, cello, flute, or piano. Making music together was part of their life. On his 60th birthday, the Philharmonic Orchestra of the Bavarian Broadcasting fulfilled his wish to play in a Mozart piano concerto with professional musicians in the original orchestration. The concert was broadcast on the radio.

High points of his domestic music making were performances of the Bach motet *Jesu meine Freude*, and the cantata *Weichet nur betrübte Schatten*, or, on Heisenberg's 65th birthday, the *Sinfonia Concertante* in E flat major and the piano concerto in D major of Wolfgang Amadeus Mozart with full orchestra in a circle of friends and students.

Einstein's Later Years: World Sage in Princeton and His "Unified Field Theory"

With his emigration to America, Einstein's international fame rose to mythical status. He became quite simply the icon of the scientist. Every child knew his face and the saucy photo in which he is sticking his tongue out at the photographer (and the world). This was partly because of the fame of the relativity theory and people's fascination with the idea of an infinite universe, and the secret of the creation of the world. In this context, Einstein appeared as Moses, climbing down the mountain bearing the tablets of the laws of stellar and planetary motion. His message is mysterious, packed up in incomprehensible mathematical formulas.

Each of Einstein's statements on whatever topic was received as indisputable wisdom. Additionally there is Einstein's ability to express complicated topics for the public in a witty or even derisive manner, in the form of *bon mots*. Many of these aphorisms became familiar quotations, such as his comparison of man's infinite stupidity to that of the infinity of the universe, or the apodictic "God doesn't play dice," with which he sought – in vain – to dismiss *ad absurdum* the insights of quantum mechanics.

He formulated this dismissal in ever changing variations, for instance in 1942, when he wrote, "It seems hard to see what cards the Lord God is holding. But I cannot for a moment believe that he plays dice and employs telepathic capabilities (as is expected of him by the current quantum theory)."

This maxim, with which he declines to accept the new reality of quantum physics, also describes the further development of his own scientific work. Einstein continued working tirelessly on his field theory, with which he attempted to unify the two "classical" theories of electrodynamics and gravitation into a single theory. He continued to fail at this task again and again. Wolfgang Pauli commented ironically on this attempt back in 1932: "Einstein's new field theory is dead. Long live Einstein's new field theory." Einstein himself conceded that his efforts had been unsuccessful. He wrote, "All my attempts to reconcile the theoretical foundations of physics with these (new) insights have totally failed. It was as if someone had pulled the rug out from under my feet, without any solid ground visible."

Still, in 1948, after Max Planck's death, he observed resignedly: "In spite of remarkable partial successes, the problem is still far from a solution."

The reason for the failure of his efforts lay partly in the fact that Einstein totally ignored all the new developments and discoveries connected with Heisenberg's quantum mechanics and Schrödinger's wave mechanics. To these belonged the discovery of the beta decay of atomic nuclei, i.e., the emission of electrons and neutrinos from the nucleus; the investigation of the strong forces between the constituents of the atomic nucleus, the protons and neutrons; and the discovery of new elementary particles in the cosmic radiation from space.

As early as 1934, Enrico Fermi published (in German) a quantum mechanical theory on beta decay: "Essay on a Theory of Beta-rays" in the *Zeitschrift für Physik*. He described this new "weak" force by analogy to electrodynamics, and thereby opened the door to a unified theory of the three fundamental forces: electrodynamics, weak and strong nuclear forces, called gauge theory. The only force which, to this day, cannot be integrated into this scheme is gravitation. Einstein's failure is understandable then. Sixty years later it has appeared that the attempt to describe gravitation in the framework of quantum mechanics, to "quantize" it, so to speak, is extremely difficult. Mathematically, it requires a sidestep into an 11-dimensional world. As a result, it is virtually impossible to derive verifiable predictions from such a theory.

The public perception of Einstein as the world sage from Princeton, who had an answer to every question of life, was not abandoned just because he had become an outsider scientifically. He carried on a massive correspondence and answered questions from students as well as from heads of state.

Fig. 6.4 Einstein in Princeton, 1946

His coequal interlocutor in Princeton was the Austrian logician, mathematician, and philosopher Kurt Gödel, with whom he often went for a walk after dinner. His second wife, Elsa, had died in 1936, one year after their move into the house at 112 Mercer Street. In 1948, his first wife, Mileva, had also passed away, in Zürich. His sister Maja who had joined him in Princeton, had died as well, in 1951. Einstein grew lonely. There remained his secretary, Helen Dukas, and the faithful Otto Nathan, both of which he appointed his executors. His relationship to his son, Hans Albert, a professor at Berkeley, and to his daughter-in-law Frieda remained strained. And his younger son Eduard had to be committed to the psychiatric hospital Burghölzli in Zürich, where his friend Carl Seelig looked after him.

Heisenberg: Government Advisor in Göttingen and Munich, Reconstruction, the Theory of Everything

Even before the war had ended, Heisenberg thought about the time following defeat. On April 28, 1944, he wrote Elisabeth, "Much is pointing to an imminent invasion. One can only hope that it comes soon." And four days later, after a long discussion with his friend, physical chemistry professor Carl-Friedrich Bonhoeffer, he wrote:

> *I greatly fear suffocating at work in the next few years. If I survive the war, I'll have perhaps ten years left in which I can hope to take an active part in science. These I would like to have totally for myself – i.e., for us – and then I will gladly thereafter fulfill my public obligations, to which the work in truth belongs. We have all lost so many years because of the war, so one has to consider exactly what one should do in the coming years. If we were in Munich, I could from time to time come to Urfeld for a few days, if I wanted to ponder things in solitude.*

This was written by the 43-year old Heisenberg, who had self-critically come to the assessment that his scientific creativity was diminishing but that he would have important work to do in the rebuilding of science following the war.

First, though, he was arrested and interned at Farm Hall with the other members of the Uranium Club. There, the German scientists learned what at first they could not believe, that the Americans had developed an atomic bomb and had used it against Japan. Heisenberg began to ponder what his future role might be. He declined offers to go to America; he intended to stay at the Kaiser-Wilhelm Institute for Physics, were he to have the opportunity.

When he was released on January 3, 1946, he went with Otto Hahn, von Laue, von Weizsäcker, and Wirtz to Göttingen in the British Zone of Occupation. The British Occupation Force was the only one to permit continuation of the Kaiser-Wilhelm Society, or its re-establishment as the Max Planck Society in its zone of occupation.

In the British zone, the liaison officer of the military authority for science, Colonel Bertie Blount, was very accommodating with respect to the wishes of the German scientists. It proved possible, through mediation of the German Scientific Council, established in January 1946, with the British military government, to re-establish the Physical-Technical Reich Institute in Braunschweig, renamed later Physikalisch-Technische Bundesanstalt, and the German Physical Society in the British zone. In October 1946, permission followed for the re-establishment of several Kaiser-Wilhelm Institutes – for physics, physical chemistry, and medical research – formerly located in Berlin, on the grounds of the former aerodynamic experimental station in Göttingen.

In February 1948, the Kaiser-Wilhelm Society became the newly established Max-Planck Society, based in Göttingen. Heisenberg saw as his tasks making the new Max-Planck Institute for Physics once again a center of experimental and theoretical physics research, contributing to the rebuilding of the largely destroyed research facilities in Germany and counseling the government.

Aside from theoretical research, the institute worked on investigating cosmic radiation and the physics of elementary particles. In the early post-war years, Heisenberg developed an approach to the theory of super-conductivity (1946-1948), a statistical theory of turbulence (1946-1948), and submitted contributions to the theory of elementary particles.

In addition to leadership of his institute, and writing his own scientific papers, Heisenberg devoted himself with great energy to the restoration of scientific research in West Germany, and especially to the realization of his ideas concerning the politics of science. On March 9, 1949, the Max-Planck Society and the West German scientific academies founded the German Research Council. Heisenberg became president of this committee, consisting of 15 scientists. The Research Council was commissioned by the federal government to represent German science in international matters. He succeeded in gaining membership for the Federal Republic of Germany in the International Union of Scientific Councils and UNESCO. Later, the German Research Foundation (DFG) arose from the union of the Research Council and the Emergency Association of German Science, established by the federal states. Here, too, Heisenberg was elected to the executive committee.

One of the most important tasks Heisenberg undertook was the restoration of international relations with German scientists. He himself accepted an invitation to lecture at British universities in 1947 and visited Niels Bohr in Copenhagen. At the negotiations leading to the foundation of the European Organization for Nuclear Research (CERN) in Geneva, beginning in 1952, Heisenberg led the German delegation. He signed the founding documents on behalf of the Federal Republic of Germany. As chair of the Scientific Policy Committee of CERN, he contributed significantly to the planning of the research program. The first general director of CERN was Heisenberg's former doctoral student at Leipzig, Felix Bloch.

Heisenberg regarded "science as an instrument for understanding among peoples," as he stressed in a talk to Göttingen students. On his initiative, the federal government established the Alexander von Humboldt Foundation, and he was named president on December 10, 1953, by Chancellor Adenauer. For more than 20 years as president, he advocated for the opportunity for young scientists from all over the world to collaborate with German colleagues at German research institutions.

When the federal republic became sovereign through the Paris accords of 1955, and joined NATO, nuclear physics research could be taken up again. In October 1955, Chancellor Adenauer set up the Federal Ministry for Nuclear Matters. Heisenberg became the chairman of the committee for nuclear physics of the Nuclear Commission. He advocated for the building of the first German research reactor by Heinz Maier-Leibnitz in Garching, outside of Munich. It began operation in October 1957.

His support for the peaceful uses of nuclear energy did not hinder him from joining other scientists in strongly opposing Chancellor Adenauer's plans to arm the German army with tactical nuclear weapons. He and other leading scientists drew up a statement against the possession and development of nuclear weapons. This declaration of the "Göttingen Eighteen", of April 12, 1957, bore the signatures of the initiators Werner Heisenberg and Carl Friedrich von Weizsäcker, and other leading physicists including Max Born, Walther Gerlach, Otto Hahn, and Max von Laue. This led to the government's abandoning its plans. At the annual gathering of Nobel laureates in Lindau on Lake Constance, Heisenberg met with his colleagues from the 1930's: Max Born, Otto Hahn, and Lise Meitner.

Fig. 6.5 Heisenberg in Lindau, 1962, with (l. to r.) Otto Hahn, Lise Meitner, and Max Born, © Werner Heisenberg Estate

Apart from all these political and administrative duties, Heisenberg devoted himself to the quest for a consistent quantum field theory of elementary particles. He hoped to find such a solution in non-linear field equations. In close collaboration with Pauli, he worked out the "non-linear spinor equation," from which he expected to be able to describe the properties of all elementary particles. Derivation of empirical consequences from the "theory of everything" proved difficult, however, and when a concrete prediction of the theory was refuted by experiments, the theory collapsed.

Heisenberg judged the new developments by American theoreticians, the quark model and the unification of the electromagnetic and the weak forces critically, although he knew the symmetries underlying the quark model and employed similar symmetries himself. These developments led to the breakthrough of today's standard model of particle physics, which was splendidly confirmed by the discoveries of the W- and Z-boson in 1984, and the Higgs-boson in 2012 at CERN. Heisenberg did not live to see these successes, which had their basis in his quantum mechanics.

The Final Meeting, 1954

The two theories, relativity theory and quantum mechanics, became the foundation of modern physics in the twentieth century. They applied and apply in various realms, relativity theory in understanding the universe and quantum mechanics in the micro-world of atoms, molecules, and elementary particles. No one has yet succeeded in unifying the two theories. Einstein wrote in March 1955 in an autobiographical memo, "It appears doubtful that a field theory can explain both the atomistic structure of matter and radiation, and also the quantum phenomena."

In the fall of 1954, Heisenberg visited Einstein at his house in Princeton. At a long conversation over coffee and cake, Einstein's entire interest was focused on the interpretation of quantum mechanics, which continued to trouble him, as it had 27 years before in Brussels. Heisenberg said that quantum theory, with its so disconcerting paradoxes, was the actual foundation of modern physics. Einstein, however, balked at granting a statistical theory such a fundamental role. "But you surely don't believe that God plays dice," said Einstein reproachfully to his guest. He could not resign himself to the idea that the reality of classical physics had been completely undermined by quantum theory. Heisenberg replied: "In quantum theory, the natural laws deal with the temporal changes of the possible and the probable. The choices that lead from the possible to the probable, however, can only be registered statistically, but can no longer be predicted."

Fig. 6.6 God plays dice, without caption, © Claus Grupen

Further development then led towards a confirmation of quantum mechanics, because experiments on the "Bell inequalities," developed by the Irish physicist John Bell at CERN, have shown that alternative theories are ruled out. Einstein´s reservations were superseded.

Einstein died in April 1955 in Princeton at the age of 76, six months after this last meeting of these two great scholars. At the memorial service, among a close circle of friends, Otto Nathan, one of the mourners, recited a verse from the Goethe's "Epilogue on Schiller's Bell," written after Schiller's death:

For he was ours!
May his proud word
Drown strongly out the noisy pain!
....
Meanwhile his soul strode strongly forth
Into the eternal true, the good, the beautiful.

Werner Heisenberg died in 1976. He was 75 years old.

The discoveries of these two geniuses will outlast the centuries.

Glossary

Action Product of work (energy) and time

Alpha particle Nucleus of the helium-atom, comprising two protons and two neutrons

Anschaulichkeit Visualizability (coined by Arthur I. Miller, MIT), Intuition, potential for visualizing, visual thinking

Atom Smallest constituent of a chemical element; consists of the positively charged atomic nucleus and the negatively charged atomic shell

Atomic nucleus Positively charged core of the atom, consisting of positively charged protons and neutral neutrons

Atomic shell negatively charged shell of the atom, consisting of electrons

Becquerel, Alexandre Edmond French physicist (1820-1891), discoverer of the photoelectric effect

Beta decay Radioactive transformation of an atomic nucleus with emission of an electron and an anti-neutrino

Beta particle Historical designation of the electron

Bethe, Hans Albrecht German-American physicist (1906-2005), studied in Frankfurt and took his doctorate in Munich with Sommerfeld. In Stuttgart, he developed a theory of the passage of fast corpuscular rays through matter. Emigration to America, Cornell University, 1938. Calculation of Carbon Fusion Cycle (CNO) in the sun, Director of the theoretical division of the Manhattan Project in Los Alamos; 1950, collaboration on the hydrogen bomb; Nobel Prize, 1967, for the Carbon Fusion Cycle

Bloch, Felix Swiss physicist (1905-1983), Heisenberg's first doctoral student and assistant in Leipzig until 1933, founder of quantum physics of solid bodies through the "Bändermodell"; 1939, U.S. citizenship; 1942-43, work on American atomic bomb project; 1954, first General Director of CERN in Geneva; Nobel Prize, 1952

K. Kleinknecht, *Einstein and Heisenberg*, Springer Biographies,
https://doi.org/10.1007/978-3-030-05264-5

Bohr, Niels Danish physicist (1885-1962), discoverer of an atomic model (1912) to explain the periodic system of the elements; patron and teacher of Heisenberg, and together with him, founder of the Copenhagen interpretation of quantum mechanics; Nobel Prize, 1922

Bohr Atomic Model Graphic representation of the atom as parallel to the planetary system, in which negatively charged electrons orbit around positively charged nucleus

Born, Max German physicist (1882-1970); teacher of Heisenberg; developed the matrix representation and the probabilistic representation of quantum mechanics; Nobel Prize, 1954

De Broglie, Louis French physicist, postulated the wave nature of the electron; Nobel Prize, 1929

Carbon Substance formed from organic material sealed off from air in geological eras; element with six protons and six neutrons in its atomic nucleus

Curie, Marie, née Sklodowska Polish/French physicist (1867-1934); discoverer of radioactivity; Nobel Prize for physics, 1903; Nobel Prize for chemistry, 1910

Diebner, Kurt German physicist (1905-1981), leader of the working group of the Army Ordnance Department within the Uranium Club for building a test reactor in Gottow

Dirac, Paul Adrien Maurice British physicist (1902-1984) after reading Heisenberg's paper of 1925 on quantum mechanics developed an alternative formulation of quantum mechanics and demonstrated the equivalence of Heisenberg's matrix mechanics with Schrödinger's wave mechanics; Nobel Prize (1933)

DPG German Physical Society, Bad Honnef; largest scientific society in the world, 50,000 members

Ehrenfest, Paul Austrian physicist (1880-1933); friend of Einstein; worked in Göttingen and St. Petersburg on statistical mechanics; professor at Leyden; suffered from depression, and committed suicide

Einstein, Albert Physicist (born 1879, Ulm, died 1995, Princeton); discoverer of relativity theory and quantum nature of light; 1902-1909, civil servant at the Swiss patent office, Bern; Professor, University of Zürich, 1909-1911; German University, Prague, 1911-1912; ETH, Zürich, 1912-1914; Prussian Academy, 1914-1933; 1934-1955, Institute for Advanced Studies, Princeton; Nobel Prize, 1922, for the explanation of the photoelectric effect

Electrodynamics Theory of the electromagnetic forces of moving charges, and of propagation of light waves

Electron Negatively charged component of the atom

Elementary Particle Fundamental building block of matter, not fissionable

Energy Universally conserved quantity in mechanical systems (work, unit Newton meter), in thermodynamic systems (heat, unit joule), and in electromagnetic systems (unit, watt seconds, kilowatt hours)

eV Electron Volt; energy an electron gains in passing through electric voltage of 1 Volt

Fermi, Enrico Italian physicist (1901-1954); theoretical and experimental physicist; worked in Pisa, Göttingen, Florence, and Rome on applications of quantum mechanics on solid bodies and quantum statistics; formulated the theory of the radioactive beta decay, and used the name neutrino for the neutral particle in beta decay postulated by Pauli; because of the peril to his Jewish wife due to the anti-Semitic laws of the Mussolini regime, emigrated, 1938, to the U.S.; 1942, built a nuclear reactor with uranium and graphite as moderator with Szilard in Chicago, and triggered the first self-sustaining chain reaction; significantly involved in the building of the American atomic bomb; Nobel Prize, 1938

Fusion Merging of hydrogen atoms to helium through high temperatures, releasing energy; energy source of the sun

Gamma radiation Energy-rich electromagnetic radiation (hard X-ray radiation); emitted through decay of excited atomic nuclei

Goudsmit, Samuel Abraham Dutch-American physicist (1902-1978); discoverer (with George Uhlenbeck) of the electron spin; from 1927, professor at Michigan; during WWII, at the radiation lab, MIT; 1945, leader of the U.S. Secret Service mission Alsos on the search for the German uranium project

Graphite Crystallized form of pure carbon, soft and dark-gray; used in pencils, electrodes, and as moderator in nuclear reactors

Gravitation Gravity; attractive force between masses; foundation of planetary motion around the sun

Gravitational Waves Waves emitted from variable masses that propagate at the speed of light, and shrink or expand space in the field of the wave

GW Gravitational wave

Hahn, Otto German chemist (1879-1968); with Fritz Strassmann, 1938, discovered through chemical analysis the fission of the heavy element uranium into two fragments; after WWII, President of the Max Planck Society; Nobel Prize for chemistry, 1944

Helmholtz, Hermann von German physicist (1821-1894); formulated the principle of conservation of energy

Hume, David Scottish philosopher and historian (1711-1776); exponent of empiricism

Isotopes heterogeneous atoms of the same chemical element, which differ only in their mass

Jordan, Pascual German physicist (1902-1980); co-author of the famous 3-person paper with Born and Heisenberg on the mathematical formulation of quantum mechanics, 1925

Kepler, Johannes German physicist and astronomer (1571-1630); court astronomer to Kaiser Rudolf II; discovered the elliptical orbits of the planets, and laws of optics

Laue, Max von German physicist (1879-1960); discovered the bending of X-rays in crystals; Nobel Prize, 1914

Lenard, Philipp German physicist (1862-1947); discovered the cathode rays and investigated the photo-electric effect; Nobel Prize, 1905; Professor at Heidelberg;

fought against relativity theory as obscure and Jewish; advocated a "German physics"

Lorentz, Hendrik Antoon Dutch physicist (1853-1928); precursor of relativity theory; inventor of the Lorentz transformations; Nobel Prize, 1902

Mach, Ernst Austrian physicist and philosopher (1838-1901); co-founder of positivism

Matrix Square of rectangular arrangement of numbers, for which the mathematical rules of a group apply

Maxwell, James Clerk Scottish physicist (1813-1879); inventor of the laws of electrodynamics, 1864

Mega- Prefix for million (M)

MeV Mega-electron-Volt: unit of energy

Micro- Prefix for millionth (μ)

Milli- Prefix for thousandth (m)

Nano- Prefix for billionth

Nuclear energy Heat energy released in the fission of an uranium nucleus

Nuclear plant Plant for production of electrical energy from uranium fission

Neutron Electrically neutral constituent of the atomic nucleus, stable in the nucleus, unstable as free particle

Nucleon Constituent of atomic nucleus; proton or neutron

Oseen, Carl Wilhelm Swedish physicist (1879-1944); formulated a theory of liquid crystals; member of the Royal Swedish Academy of Sciences from 1921, and a member of the Academy's Nobel Prize committee for physics from 1922.

Photon Particle of light; light quantum

Photo-voltaic Generation of electric potential directly from sunlight with solar cells of silicon or gallium arsenide

Planck, Max German physicist (1858-1947); explained the colors (frequencies) of the radiation of a hot black body by the postulate that energy can only be exchanged in the smallest packets, quanta; introduced the quantum of action, Planck's Constant h

Pauli, Wolfgang Austrian physicist (1900-1958), student friend of Heisenberg; formulated the exclusionary principle for electrons in the atom; Nobel Prize, 1945

Proton Stable, positively charged building block of the atomic nucleus

Quantum Mechanics Theory of the processes in the atomic realm; first formulated as matrix mechanics by Heisenberg, 1925, then by Schrödinger as wave mechanics, 1926

Quantum Jump Transition of an atom from a state with a certain energy to another state by emission or absorption of a light quantum

Quantum of Action Smallest unit of action, called by Max Planck the quantum h

Quantum of Action $\hbar$ Symbol for the expression $h/(2\,\pi)$

Quantum Number In the Bohr model, the Quantum states of an electron in the atom are ordered by three natural numbers, corresponding to the energy and the angular momentum of the electron; a fourth half-integral quantum number (introduced by Pauli) corresponds to the intrinsic angular momentum of the electron

Radioactivity Transformation of an atomic nucleus with emission of radiation or charged elementary particles

Radium Unstable element with the atomic number 88, that emits radioactive radiation during its decay

Reactor Machine for producing heat and electrical energy from nuclear fission

Schrödinger, Erwin Austrian physicist (1887-1961); after Heisenberg, in 1926 invented a second form of quantum mechanics: wave mechanics; the Schrödinger equation is more easily manageable than matrix mechanics; Nobel Prize, 1933

Siemens, Werner von German physicist and engineer (1816-1892), inventor of the electrical generator

Sommerfeld, Arnold German physicist (1868-1951); enlarged and improved significantly Bohr's model of the atom; teacher of Pauli and Heisenberg

Spectral line Monochromatic light of a color that arises from a quantum jump in an atom

Spin Intrinsic angular momentum of an elementary particle that can be oriented in two directions in a magnetic field

Swiss Polytechnic, Zürich Technical university supported by the Swiss federal government; founded, 1855; right to award doctorate from 1908 onwards; renamed ETH (Eidgenössische Technische Hochschule), 1911

Szilard, Leo Hungarian physicist (1898-1964); studied physics from 1919 in Berlin, and took his doctorate in 1922 on a thermodynamic subject. 1933, emigrated to England, 1934, idea of a chain reaction in nuclear transformations; 1938 to the U.S.; building of the first uranium reactor with Fermi, and first self-sustaining chain reaction; moving force behind the American bomb project; persuaded Einstein to write letter to Roosevelt for building an atomic bomb, 1939; tried in vain to prevent dropping of the bomb on Japan, 1945, by signing the Franck Report together with 70 other scientists

Teller, Eduard (Edward) Hungarian physicist (1908-2003); studied at the Technical University, Karlsruhe, and took his doctorate under Heisenberg in Leipzig; emigrated in 1933 via England to the U.S.; worked at Los Alamos on the atomic bomb, and after 1950 promoted development of the hydrogen bomb

Uncertainty In the atomic realm, position and speed of a particle cannot simultaneously be determined exactly; the product of the uncertainties of the two values is limited by Planck's constant, according to Heisenberg

Uranium Heavy element with 92 protons and 92 electrons in an atom; the element with 146 neutrons in its nucleus (uranium 238) is stable; the element with 143 neutrons (uranium 235) can be split by slow neutrons, and is the fuel of nuclear reactors

Volt Unit of electrical potential

Volta, Alessandro Italian physicist (1745-1827)

Watt Unit of electrical power; a voltage source that produces a current of 1 ampere and 1 volt provides 1 watt; 1,000 watts correspond to 1.36 horsepower

Watt, James English physicist (1736-1819); inventor of the steam engine

Weizsäcker, Carl-Friedrich von German philosopher and physicist (1912-2007); studied with Heisenberg and Friedrich Hund; explained the carbon-fusion cycle in the sun, independent of Hans A. Bethe; worked from 1940 to 1942 as theorist on the German uranium project; professor at the Reichsuniversität Straßburg in occupied France, 1942; after 1945, oriented to philosophy and politics

Wien, Wilhelm German physicist (1864-1928); inventor of the Wien Displacement Law of heat radiation; Nobel Prize, 1911

X-Radiation Energy-rich, invisible light radiation that penetrates body tissue, and serves to make bones and the lung visible; discovered in 1895 by Conrad Röntgen in Würzburg

Zeeman Effect Splitting of spectral lines of an atom in a magnetic field

Zeeman, Pieter Dutch physicist (1865-1943); discovered the splitting of spectral lines of luminous matter in an external magnetic field; Nobel Prize, 1902

Bibliography

Literature on Einstein

Lorenz Friedrich Beck, *Max Planck im Kaiserreich und in der Weimarer Republik,* in *Max Planck und die Max-Planck-Gesellschaft* (Max Planck in Imperial Germany and during the Weimar Republic), ed. M. Kazemi, Berlin 2008, Max-Planck-Gesellschaft.

Alice Calaprice, Daniel Kennefick, Robert Schulmann, Einstein Encyclopedia, Princeton University Press, 2015

Albert Einstein, *Aus meinen späten Jahren* (From My Late Years), Deutsche Verlagsanstalt, Stuttgart, 1984.

Albert Einstein/Mileva Maric: *The Love Letters,* ed. J. Renn, R. Schulmann, Princeton University Press, 2000; (German: Am Sonntag küss ich Dich mündlich, Die Liebesbriefe 1897–1903, Piper Verlag München Zürich, 1998).

Philipp Frank, *Albert Einstein, Sein Leben und seine Zeit* (Albert Einstein, His Life and Science), Mit einem Vorwort von (With forward by) Albert Einstein 1942, Paul List Verlag München, 1949.

Siegfried Grundmann, *Einsteins Akte, Wissenschaft und Politik – Einsteins Berliner Zeit* (Einstein´s Dossier, Science and Politics – Einstein´s Berlin Period), Springer Berlin-Heidelberg, 2004.

Armin Hermann, *Einstein,* Piper München, 1994.

Roger Highfield und Paul Carter, *The Private Lives of Albert Einstein,* St. Martin's Griffin, 1994; (German: Die geheimen Leben des Albert Einstein, Marix Verlag Wiesbaden, 2004).

Max Jammer, *Einstein and Religion,* Princeton University Press, 1999.

Robert Jungk, *Brighter Than a Thousand Suns; a Personal History of the Atomic Scientists,* New York, 1958. (German: Heller als tausend Sonnen, Rowohlt Verlag, Hamburg, 1956).

K. Kleinknecht, *Einstein and Heisenberg,* Springer Biographies,
https://doi.org/10.1007/978-3-030-05264-5

Abraham Pais, *Subtle is the Lord, The Science and the Life of Albert Einstein,* Oxford University Press, 1982.

Richard Rhodes, *The Making of the Atomic Bomb,* Simon and Schuster, New York, 1986.

Jamie Sayen, *Einstein in America,* Crown Publishers, New York, 1985.

Alexis Schwarzenbach, *Das verschmähte Genie, Albert Einstein und die. Schweiz* (The Disdained Genius, Albert Einstein and Switzerland), DVA München, 2005.

Literature on Heisenberg

Hans A. Bethe, "The German Uranium Project," *Physics Today* Online, Vol. 53, No.7 (July 2000)

Barbara Blum-Heisenberg, *Werner Heisenberg und die Musik – ein anderer Zugang zu meinem Vater,* (Werner Heisenberg and His Music – Another Approach to My Father), Privatschrift.

Gerd W. Buschhorn und Helmut Rechenberg, *Werner Heisenberg auf Helgoland* (Werner Heisenberg on Helgoland Island), Max-Planck-Institut für Physik (Werner-Heisenberg-Institut), München 2000.

Cathryn Carson, *Heisenberg in the Atomic Age* (Cambridge University Press, 2010)

David C. Cassidy, *Werner Heisenberg,* Leben und Werk (Spektrum Verlag, 1995).

David C. Cassidy, *Beyond Uncertainty* (Bellevue Literary Press, 2009).

Jérome Ferrari, *Das Prinzip – wie Werner Heisenberg uns zeigte, dass uns mit dem Schönen die Welt verloren geht* (The Principle – How Werner Heisenberg Showed Us That By Losing Beauty We Lose the World), Secession Verlag Zürich 2015.

Ernst Peter Fischer, *Werner Heisenberg – ein Wanderer zwischen zwei Welten* (Werner Heisenberg – A Wanderer Between Two Worlds), Springer Heidelberg 2015.

Ernst Peter Fischer, *Werner Heisenberg – Das selbstvergessene Genie* (Werner Heisenberg, the Oblivious Genius), München, 2001.

Klaus Gottstein, *Heisenberg and the German Uranium Project* (1939–1945). Myths and Facts (Juni 2016), http://www.heisenberg-gesellschaft.de.

Samuel A. Goudsmit, *Alsos,* American Institute of Physics, Woodbury, New York, 1996

Elisabeth Heisenberg, *Inner Exile: Recollections of a Life with Werner Heisenberg,* Birkhäuser Verlag Basel, 1984, (German: Das politische Leben eines Unpolitischen, Piper München, 1983).

Werner Heisenberg, *Physics and Beyond; Encounters and Conversations,* New York, 1971a. (German: Der Teil und das Ganze, Gespräche im Umkreis der tomphysik, Piper München, 1969)

Werner Heisenberg, *Schritte über Grenzen, Gesammelte Reden und Aufsätze* (Steps Over Borders, Lectures and Essays), Piper München, 1971b.

Werner Heisenberg, *Ordnung der Wirklichkeit* (Order of Reality), Piper München, 1989.

Werner Heisenberg, *Tradition in der Wissenschaft* (Tradition in Science), Piper München, 1977.

Werner Heisenberg, *Liebe Eltern! Briefe aus kritischer Zeit* (Dear Parents! Letters from Critical Times), 1918 bis 1945, Hg. Anna Maria Hirsch-Heisenberg, Langen Müller München, 2003.

Werner Heisenberg, *Elisabeth Heisenberg, My Dear Li, Correspondence 1937-1946,* ed. Anna Maria Hirsch-Heisenberg, Yale University Press, 2016. (German: Meine liebe Li, Der Briefwechsel 1937–1946, Residenz Verlag St. Pölten-Salzburg, 2011)

Werner Heisenberg, *Gutachten- und Prüfungsprotokolle* (Advisory Reports and Records of Exams), 1929–1942, herausgegeben von H. Rechenberg und G. Wiemers, ERS Verlag Berlin, 2001.

Werner Heisenberg, *Gesammelte Werke, herausgegeben von Walter Blum, Hans-Peter Dürr und Helmut Rechenberg: Abt. A: Wissenschaftliche Originalarbeiten, Abt. B: Wissenschaftliche Übersichtsartikel* (Collected Works, Sect. A: Original Scientific Papers, Sect. B: Review articles, lectures, books, Sect. C: writings for the general public), Vorträge, Bücher, Springer Verlag Berlin und Heidelberg (1984 ff.) Abt. C: Allgemeinverständliche Schriften, C I bis C V, Piper Verlag München (1984 ff.).

Armin Hermann, *Werner Heisenberg* (rororo, Hamburg, 1976).

Arthur I. Miller, *Imagery in Scientific Thought Creating 20th-Century Physics,* The MIT Press, Cambridge, Massachusetts and London, England, 1986; reprint 2017 Springer Nature, Switzerland

Thomas Powers, *Heisenberg's War: The Secret History of the German Bomb,* Little, Brown & Company, 1993.

Helmut Rechenberg, *Werner Heisenberg – Die Sprache der Atome* (Werner Heisenberg, the Language of the Atoms), Band 1 and 2 (Springer, Heidelberg, 2010).

Michael Schaaf, *Heisenberg, Hitler und die Bombe, Gespräche mit Zeitzeugen* (Heisenberg, Hitler and the Bomb, Interviews with Contemporary Witnesses), GNT-Verlag, Berlin-Diepholz (2001).

Gregor Schiemann, *Werner Heisenberg* (C. H. Beck, München, 2008).

Richard von Schirach, *Die Nacht der Physiker* (The Night of the Physicists), Berenberg Verlag Berlin 2001.

Edward Teller, "Heisenberg Sabotaged the Atomic Bomb," Interview (1995) with Michael Schaaf, p.114 in Schaaf (2001).

Author Index

K. Kleinknecht, *Einstein and Heisenberg*, Springer Biographies,
https://doi.org/10.1007/978-3-030-05264-5

Subject Index

K. Kleinknecht, *Einstein and Heisenberg*, Springer Biographies,
https://doi.org/10.1007/978-3-030-05264-5

Picture Credits

Chapter 1: "Einstein's Youth"

- Fig. 1.1: Headstones at the Jewish cemetery of Buchau, © Konrad Kleinknecht, Page 2
- Fig. 1.2: Hermann and Pauline Einstein, Wikimedia Commons, Page 3
- Fig. 1.3: Albert Einstein with his sister Maria (Maja), 1885, © Bildarchiv Preußischer Kulturbesitz (PBK), Page 4
- Fig. 1.4: Dwelling of the Einstein family on Adlzreiterstraße, Munich, © Konrad Kleinknecht, Page 6
- Fig. 1.5: Gymnasium student Albert Einstein, age 14 years, in Munich, © PBK, Page 7
- Fig. 1.6: Mileva Maric' in Zürich, 1900, © PBK, Page 11
- Fig. 1.7: Einstein at the patent office, Bern, 1905, © PBK, Page 14
- Fig. 1.8: Albert and Mileva Einstein in Bern, 1905, © PBK, Page 16

Chapter 2: "Heisenberg's Youth"

- Fig. 2.1: The family of August and Annie Heisenberg, with their sons Erwin and Werner, on a hike near Würzburg, 1907, © Werner Heisenberg Estate, Page 19
- Fig. 2.2: Werner (left) and Erwin visiting their grandfather in Osnabrück, 1906, © Werner Heisenberg Estate, Page 20

K. Kleinknecht, *Einstein and Heisenberg*, Springer Biographies,
https://doi.org/10.1007/978-3-030-05264-5

- Fig. 2.3: Nikolaus Wecklein with his grandsons, Erwin and Werner, 1915, © Werner Heisenberg Estate, Page 21
- Fig. 2.4: Arnold Sommerfeld, 1920, © Werner Heisenberg Estate, Page 25
- Fig. 2.5: Wolfgang Pauli 1918, © CERN, Geneva, Page 26
- Fig. 2.6: Heisenberg's Göttingen teachers, Max Born (seated), and James Franck (2nd from right), with visitors Carl Wilhelm Oseen (left), Niels Bohr (2nd from left) and Oskar Klein , scientific assistant to Bohr (right), June 1922, © Werner Heisenberg Estate, Page 31

Chapter 3: "The Wonder Years"

- Fig. 3.1: Albert and Mileva Einstein in Zürich, 1910, © PBK, Page 49
- Fig. 3.2: Berliner Illustrirte 1919: A new great figure of world history, Wikimedia Commons, Page 59
- Fig. 3.3: Einstein in the library of the Kaiser-Wilhelm-Institute for physics 1921, © PBK, Page 60
- Fig. 3.4: Heisenberg, 1925, © Werner Heisenberg Estate, Page 65
- Fig. 3.5a: Manuscript 1 on the uncertainty principle, © Werner Heisenberg Estate, Page 76
- Fig. 3.5b: Manuscript 2 on the uncertainty principle, © Werner Heisenberg Estate, Page 77
- Fig. 3.6: Uncertainty © Claus Grupen, Page 79
- Fig. 3.7: Pauli, Heisenberg, and Fermi at the Como-Conference, 1927, © CERN, Geneva, Page 80

Chapter 4: "Impact of the Discoveries"

- Fig. 4.1: Participants of the Solvay Conference, 1927, © CERN, Geneva, Page 82
- Fig. 4.2: House of famous visitors © Claus Grupen, Page 85
- Fig. 4.3: Einstein in a Black Hole © Claus Grupen, Page 91
- Fig. 4.4: Signal of gravitational wave GW150914, LIGO collaboration (left); formation of the gravitational wave GW150914 (right); "Strain" signifies the relative change in length of the spectrometer arms during the passage of the gravitational wave, LIGO collaboration, B. Barish (Caltech), Page 94
- Fig. 4.5: Heisenberg at his introductory lecture in Leipzig, 1928, © Werner Heisenberg Estate, Page 95

- Fig. 4.6: Heisenberg with his students in 1930. Left to right: Giovanni Gentile, Rudolf Peierls (front), George Placzek, Gian Carlo Wick, Werner Heisenberg (front), Felix Bloch, Viktor Weisskopf, Fritz Sauter, © Werner Heisenberg Estate, Page 97
- Fig. 4.7: Skiing with Felix Bloch, 1933, © Werner Heisenberg Estate, Page 99
- Fig. 4.8: Elisabeth and Werner Heisenberg, 1937, © Werner Heisenberg Estate, Page 101
- Fig. 4.9: The new picture of the atom, archive of the author, Page 105
- Fig. 4.10: One of many applications: the MRI process in medicine, Wikimedia Commons, Page 109

Chapter 5: "Expulsion and the War Years"

- Fig. 5.1: Einstein with Nernst, Planck, Millikan, and v. Laue, 1931, © PBK, Page 113
- Fig. 5.2: Homeward bound from Japan with Elsa, 1922, © PBK, Page 114
- Fig. 5.3: Einstein's letter to Dr. Dielmann, 1952, Schlossmuseum Büdingen, Page 116
- Fig. 5.4: Poster of the appeal to the SPD and KPD, archive of the author, Page 119
- Fig. 5.5: Factory for the separation of isotopes, Oak Ridge, Wikimedia Commons, Page 126

Chapter 6: "Social Affinities"

- Fig. 6.1: Einstein with Elsa and step-daughter Margot, 1927, © PBK, Page 144
- Fig. 6.2: Einstein in the tower room of 5 Haberlandstraße, Berlin, 1927, © PBK, Page 147
- Fig. 6.3: The Measuring God, Bible Moralisée, 1250, Reims, Archive of the author, Page 151
- Fig. 6.4: Einstein in Princeton, 1946, © PBK, Page 158
- Fig. 6.5: Heisenberg in Lindau, 1962, with (l. to r.) Otto Hahn, Lise Meitner, and Max Born,© Werner Heisenberg Estate, Page 162
- Fig. 6.6: God plays dice, © Claus Grupen, Page 163

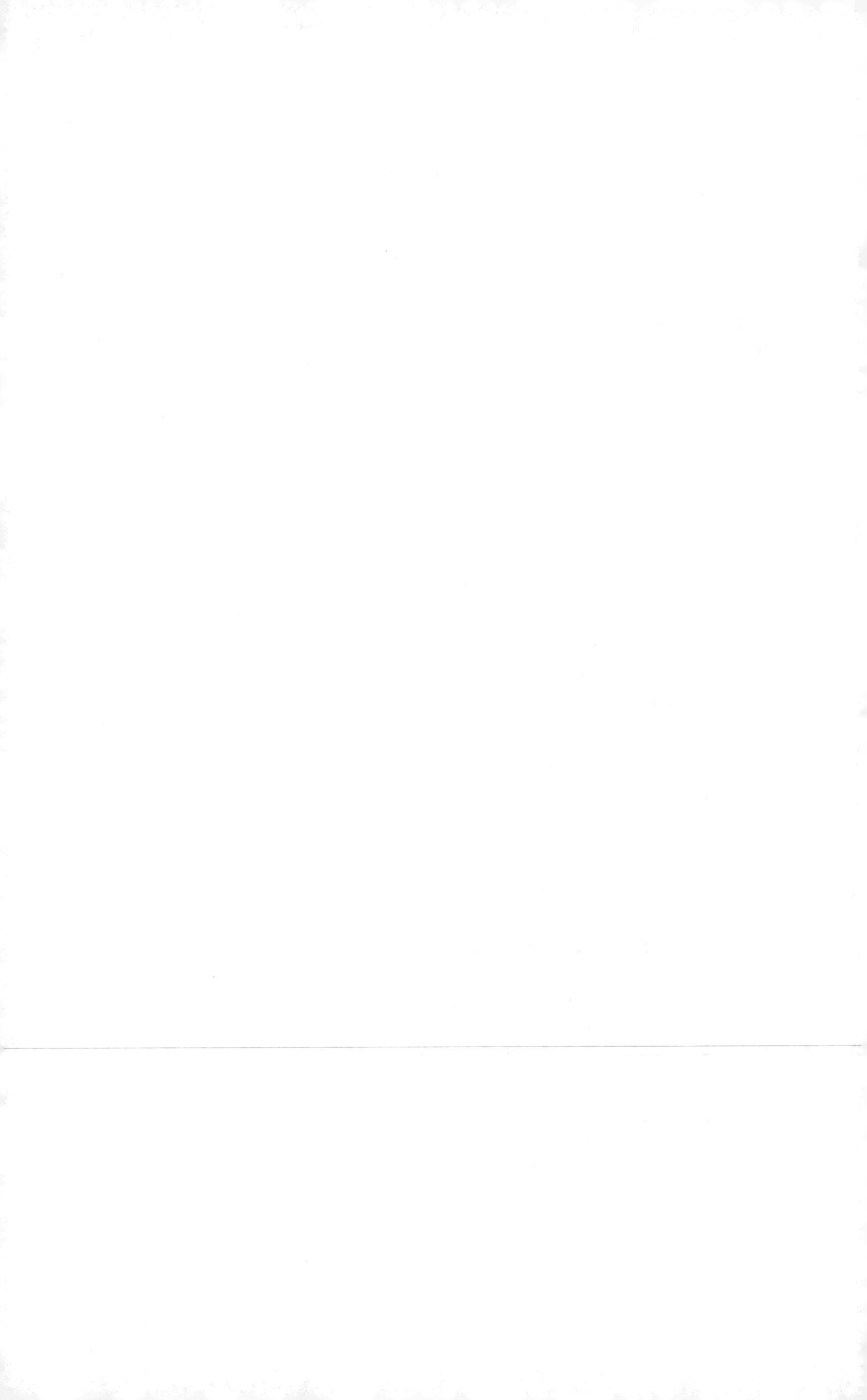